The Tangled Bank

Princeton Theological Monograph Series

K. C. Hanson, Charles M. Collier,
and D. Christopher Spinks, Series Editors

Recent volumes in the series:

Kevin Twain Lowery
*Salvaging Wesley's Agenda:
A New Paradigm for Wesleyan Virtue Ethics*

Ralph M. Wiltgen
*The Founding of the Roman Catholic Church
in Melanesia and Micronesia, 1850–1875*

Steven B. Sherman
*Revitalizing Theological Epistemology:
Holistic Evangelical Approaches to the Knowledge of God*

Matthew J. Marohl
Faithfulness and the Purpose of Hebrews

Stanley D. Walters, editor
Go Figure! Figuration in Biblical Interpretation

Paul S. Chung
Martin Luther and Buddhism: Aesthetics of Suffering, Second Edition

D. Seiple and Frederick W. Weidmann, editors
*Enigmas and Powers: Engaging the Work of Walter Wink
for Classroom, Church, and World*

Mary Clark Moschella
*Living Devotions: Reflections on Immigration,
Identity, and Religious Imagination*

The Tangled Bank
*Toward an Ecotheological Ethics
of Responsible Participation*

MICHAEL S. HOGUE

☙PICKWICK *Publications* • Eugene, Oregon

THE TANGLED BANK
Toward an Ecotheological Ethics of Responsible Participation

Princeton Theological Monograph Series 89

Copyright © 2008 Michael S. Hogue. All rights reserved. Except for brief quotations in critical articles or reviews, no part of this book may be reproduced in any manner without prior written permission from the publisher. Write: Permissions, Wipf and Stock Publishers, 199 W. 8th Ave., Suite 3, Eugene, OR 97401.

Pickwick Publications
A Division of Wipf and Stock Publishers
199 W. 8th Ave., Suite 3
Eugene, OR 97401

www.wipfandstock.com

ISBN 13: 978-1-55635-380-2

Cataloging-in-Publication data:

Hogue, Michael S.

    The tangled bank : toward an ecotheological ethics of responsible participation / Michael S. Hogue.

    xxxii + 250 p. ; 23 cm. — Includes bibliographical references.

    Princeton Theological Monograph Series 89

    ISBN 13: 978-1-55635-380-2

    1. Human ecology—Religious aspects—Christianity. 2. Nature—Religious aspects—Christianity. 3. Environmental ethics. 4. Human beings—Effect of environment on—Moral and ethical aspects. 5. Nature—Effect of human beings on—Moral and ethical aspects. 6. Jonas, Hans. 7. Gustafson, James M. I. Title. II. Series.

BT695.5 H65 2008

Manufactured in the U.S.A.

To my parents,
Donald Gary and Karen Hogue

Contents

Preface ix

Introduction xiii

PART ONE: Hermeneutical Dimensions

1 Human Power and Existential Biology 3

2 Divine Power and Critical Religious Naturalism 28

PART TWO: Fundamental Dimensions

3 Technos, Bios, Anthropos 65

4 God, World, Human Being 98

PART THREE: Normative Dimensions

5 The Imperative of Responsibility 137

6 Theocentric Ethical Participation 184

Conclusion: Toward an Ecotheological Ethics of Responsible Participation 225

Bibliography 247

Preface

GROWING UP THE SON OF A PREACHER IN THE WOODS AND LAKES OF northwest Michigan (where, if you hold up your Michigan-shaped left hand, your pinky touches the ring finger), nature and religion have always been prominent in the ecology of my life. But they were often separate. My religious life was for a long time a largely private affair, the personal business of inwardness. It was what I attended to on Sundays and occasionally in independent reading and prayer. My life in nature, on the other hand, was the rest of the rest. I spent much of my childhood immersed in the beauties of the Great Lakes region, often water-skiing from morning to evening on the crystal cold lakes that pockmark northern Michigan, frequently camping in its many square miles of woods, and sometimes getting lost and found while hiking and biking in the network of trails on the outskirts of my hometown. While growing up, my work life was also rooted in the natural world. I was variously employed on cherry farms, on the beaches of Lake Michigan, by an environmental newspaper, and by a land conservancy to build trails on its land preserves.

I have changed a great deal in the many years that have passed since I moved away from northern Michigan. I have experienced more of the worlds of nature and of religious life. I have traveled through and lived in many wilderness and urban places, appreciating their many beauties and also mourning their many wounds. And as a religious person, my faith has emerged into something much more public and engaged than it used to be. I remain the religious inquirer that I always have been, but I have also transitioned from being a student of theological ethics at a divinity school to a theological educator at a seminary.

In the time between my childhood and adulthood, as well as between the time when I began this book and the time that I completed it, I have worked more deliberately to bring together my love and concern for the natural world with my commitments as a religious person and theological scholar. In this more deliberate work, of which this book has been part, I have come to recognize the degree to which all along, even and especially when I didn't recognize it, the various stages and places and natural and religious habitats of my life have always intermingled. I have come to recognize in a biographical sense what John Muir once said of the universe, "When we try to pick out anything by itself, we find it hitched to everything else . . . ."

I am not exactly the same person I was when I was growing up in northern Michigan. Nor am I exactly the same person today that I was when I began writing this book. If I were to start over on this book today, it would likely end up being quite a bit different. Yet the different selves I have been in the different places and times of my life are all a part of who I am now and what this book has become. And so while it is not at all explicit, there is a little bit of northern Michigan beneath the surface of this book. Deep within the scholarship, between the lines, there is a little bit of northern Michigan's clear night skies and a little of its rain that comes like a veil across the lake. And within my conversations with Hans Jonas and James M. Gustafson, around which this book is organized, there is the influence of the many lives of the people I have known in northern Michigan, their hardships as farmers in bad times and their inspiring efforts to preserve the beauty of the land.

All this is to say that in the course of writing this book, from the earliest stages of research to the final phases of editing, I have learned how true to life Charles Darwin's image of the "tangled bank" is. No life is ever fully independent of other lives, and my life during the time that I have been writing this book has intersected with and been nourished by many others. I hope that you will read what I have to offer here as an expression of and an invitation to ongoing collaboration on one of the most important tasks of our time, responsibly participating as religious and theological scholars in the work of preserving the possibility and even nourishing the futures of nature and human life.

In keeping with my book's title, my community of support in the several years of writing has been very much like a tangled bank. My research would neither have had a beginning nor come to a conclusion (open as it remains), without a sustaining network of reinforcing interdependencies. Professor William Schweiker, my former doctoral advisor at the University of Chicago, has been of especially momentous help. He is a master scholar and a gifted teacher-mentor, a true Socratic midwife. His encouragement and expertise have been invaluable to me, and I am honored now to know him and to continue to work with him as a friend and colleague.

I am also fortunate to have a mother and father who not only understand and support my scholarly vocation, but who also read multiple drafts of my chapters and provided helpful technical and substantive contributions along the way. We have spent many late evenings talking through the frustrations and epiphanies that invariably attend the writing of a book. My father especially has worked with me through nearly every detail of my work. He stands for me as a great example of religiously faithful and morally engaged intellectual life.

I would also like to thank Kincade, my young son, who was barely walking when I began this work and is now a soccer player, a baseball slugger, and on occasion, a cleaner of his bedroom. His smiles and laughter and joy for life, and his troubles and his learning, inspire my own life and remind me continually of the larger purposes of intellectual and religious life. I have learned more about the meaning and importance of responsible participation as his father than from any of the many books that line my shelves.

And lastly, and most importantly, I want to thank my wife, Sara, for truly making everything better.

Introduction

WITH TYPICAL VICTORIAN RESERVE, CHARLES DARWIN WRITES IN THE concluding paragraph of his revolutionary work, *The Origin of Species*,

> It is *interesting* to contemplate *a tangled bank*, clothed with many plants of many kinds, with birds singing on the branches, with various insects flitting about, and with worms crawling through the damp earth, and to reflect that these elaborately constructed forms, so different from each other, and dependent upon each other in so complex a manner, have all been produced by laws acting around us.[1]

But the biotic diversity and the interrelation among species and their habitats Darwin perceives are not simply interesting. Even on the tiny scale of some nameless patch of earth, this diversity of interrelations is shattering! Adding to Darwin's awe is his conviction that some set of laws underlies, produces, and sustains the variety of living forms, their difference from as well as their dependence on one another. Biotic diversity—the origin, proliferation, dying out, and intermingling of species—is constituted by the pressure of natural selection acting upon the organic forces of growth, reproduction, inheritance, and variation. From the dynamism of this pressure and these forces, "from the war of nature, from famine and death, the most exalted object which we are capable of conceiving, namely, the production of the higher mammals, directly follows."[2]

As the production of higher mammals follows from this dynamism, so also do human beings. But with humans come many capacities that seem unrelated to this story of evolution, perhaps even contrary to it. For humans are moral and religious animals as well, capable of conceiving and choosing the right, of envisioning and acting toward the good, of registering and constructing visions of the ultimate purposes

---

1. Darwin, *The Origin of Species*, 459 (my italics).
2. Ibid.

of existence. While often selfishly pursuing our own interests, on many occasions we sacrifice our individual and immediate concerns for the interests of other individuals or for the good of a community. While we certainly make war, while many become rich through the exploitation of others, we also make peace and strive for justice.

How can a creature capable of such morally worthy projects be the product of nature's "war," a war in which to sacrifice oneself and through oneself one's offspring seems to run counter to the central biotic drive of survival? How can a morally striving religious animal be the product of what seems at best to be the amoral nature of mundane evolutionary processes? Darwin's picture of the biotic diversity and the interrelationships among species in their riparian habitat provokes interesting questions indeed, especially questions about how to understand the meaning and responsibilities of the human moral relation to the natural world!

Darwin's "tangled bank" of interdependency provides a suggestive framing image for the intertwined concerns of this book. It indicates the complexity of exploring the meaning and moral responsibilities of being human through the relations among the biological sciences, ethics, and theology. As with the multiple forms of life inhabiting Darwin's bank, biological theory, ethics, and theology are not entirely autonomous from one another. While different, their boundaries are porous rather than impermeable. Biological theory, ethics, and theology are matrices of inquiry that significantly impinge on one another.

Among its other tasks, this book examines the interstices within the "tangled bank" of these disciplines in order to gain a better view of the moral meanings and responsibilities of the human as a natural historical creature. Looking into these questions is especially important, I argue, in a time of environmental crisis. And so more constructively, the image of the "tangled bank" also provides a metaphor for an eco-theological ethics of responsible participation, which this book proposes as a promising direction for an ethics for our time. In addition, "the tangled bank" provides a rhetorical framing motif for comparative critique of Jewish philosopher Hans Jonas and Christian theologian James M. Gustafson, thinkers who have been under-appreciated within the field of environmental ethics but that I will advance as significant resources. And last, Darwin's "tangled bank" points toward what I take to be one of the most pressing environmental ethical problems of our time—the conjunction of radical power and moral uncertainty.

Power itself is not the problem, for power is merely the potential to impact the world in some way. Power harbors as much a constructive potential for good as a destructive potential for harm. The problem with power is knowledge of what to do with it—how to apply it, what ends to direct it toward, how to constrain it. Whether the impact of power on the world is for good or evil turns on the freedom of individuals, communities, and societies to choose how to exercise power. Choice is entailed in the application of power. And the possibility of choice assumes that those choosing are sufficiently free to imagine and actually to bring about the alternative outcomes that are the subject of choice. Choice is correlative to power whether power is related existentially to human individuals or socially to communities. Whether understood in a principally negative way as domination, on a more positive collaborative model, or in terms of some other construction, power and choice go hand in hand and always will.

But in our time this correlation has taken on a new moral gravity. For at the very moment when the capacities of our individual and communal power to alter the world are so profoundly increasing, as individuals and communities we are less morally confident than ever about how to exercise that power. One of the most acute moral problems for us, then, is that the colossal growth of our human power to impact the world is joined by an increasing uncertainty about how morally to guide this power. This book analyzes this relatively undertheorized problem in relation to the environmental crisis and technological culture, drawing upon the insights of Jonas and Gustafson and bringing them forward as crucial but neglected resources for constructive thinking in environmental ethics.

The rise of moral uncertainty has many sources. In at least two ways, the growth of human power itself is one of these. New scientific and technological developments continuously reframe our understandings of the context and stakes of moral action. As we learn more from the natural sciences about our world we come to a greater understanding of the complex impacts of human behavior in it. Awareness of the interdependencies of our world challenges our frameworks for determining the consequences of action, for weighing risks and opportunities, for determining what is right and good. In all of these ways and more, increase of knowledge and power complicates the moral life. So the growth of power through the combination of new knowledge and technologies is a generative cause of our moral uncertainty.

A second cause of the rise of moral uncertainty stems from our increasing awareness of and sensitivity to moral pluralism. This too is related to the expansion of human power. For our awareness of moral diversity is in part the effect of innovative communication and transportation technologies that bring the far reaches of the world closer to home, fomenting contact and exchange among different cultures in unprecedented ways.

While increasing knowledge of the vast differences that structure our world may in essence be a positive good, one of its negative effects can be the protraction of ethical and policy debates regarding the exercise of power. Debate over important matters and consideration of moral differences are properly a part of the process of careful moral deliberation and a necessary prelude to thoughtful action. And yet when considered debate devolves into reticence, quietude, or paralysis we fail morally in the face of the gravity of our expanded power. Power always entails that moral choices must be made and the significance of choice in our time of radical power is profoundly magnified. So a significant task for contemporary ethics, philosophical and theological, environmental and otherwise, entails facing the problematic conjunction of radical power and moral uncertainty.

Another source of our moral uncertainty is the rise of historical consciousness. Philosophers, cultural theorists, anthropologists and others have in recent decades severely called into question the characteristic modern Enlightenment search for absolutes, for a neutral epistemic foundation from which to access objective truth. The understanding that all quests for knowledge and norms are in some way interested, perspectival, historically situated, culturally partial, and socially particular is now exceedingly familiar. Assumptions about the ahistorical nature of reason have ceded to emphases on contingency. We know that knowers are not only passive receptacles of the real but participate in the construction of reality. The quest for universally binding moral norms has largely given way to descriptions of the cultural and historical embeddedness of morality.

The rise of historical consciousness thus raises the specters of epistemic and moral relativism. Nostalgic calls for a return to an ahistorical vision of reason and morality are certainly out of place. And yet it is equally certain that our moral uncertainty is a grave problem in the face of our expanding powers to alter the world, and specifically the natural environment that sustains us as well as all other forms of life.

This book is motivated by the problem of power and uncertainty as one of the most pressing moral problems of our time. But this is a vast problem and responding adequately to it requires choosing a point of entry, isolating a topic that enables seeing it in finer-grained detail. So while generally concerned with the problem of power and moral uncertainty, this book is more specifically concerned with power and uncertainty in light of our environmental crisis and technological culture. My aim, in light of these specific concerns, is to articulate the contours of an ecotheological ethics of responsible participation, utilizing a moral anthropological method to comparatively critique the intersection of biological theory, ethics, and theology in the works of Jonas and Gustafson.

I aim to do several things in this introduction. I will first describe the roles of power and moral uncertainty in the environmental crisis. Following this, I will propose a moral anthropological method as a suggestive but underdeveloped strategic option for theory construction and interpretation in environmental ethics. I will then present the basic shape of my argument and introduce the ethical theories of Jonas and Gustafson.

Power, Uncertainty, and Environmental Crisis

The conjunction of radical power and moral uncertainty manifests itself in multiple interrelated aspects of contemporary moral experience, existential, political, and social, each of which warrants careful consideration. Yet underlying and reaching through each of these manifestations is one that I take to be in some ways more fundamental—the environmental crisis. The expansion of human power through the invention of new technologies, and the swirling debates regarding choices and policies about the guidance of this power, have introduced unprecedented moral dilemmas and confusion concerning the human relationship to the natural world.

Never before has humanity had the knowledge, skill, tools, or power irreversibly to alter the natural environment surrounding and sustaining it. Nature before recent times was a reliably stable backdrop to human existence. While there never was a time in human history when humans did not impact nature, the range of this inevitable human impact was until the last couple of centuries relatively limited. There is an enormously wide moral gap between the capacity to change nature partially and the capacity to alter it totally. Navigating this gap is one of the most significant ethical tasks of our time.[3]

The environmental crisis can be understood as a fundamental moral problem for at least two reasons. First, insofar as the existential, social, and political dimensions of moral experience depend upon an ecological order sufficient to sustain them, then an environmental crisis that may compromise that order is more basic. Both the moral experience of individuals and the moral structure of societies simultaneously rest upon and are shaped by the condition of natural environments. Natural environments are fundamental to human moral experience, then, in that they provide a framework that shapes the limits and possibilities of such experience. This is not to say that natural environments are entirely determinative of human experience, for historical and cultural contexts also shape experience in profound ways. Yet, while history and culture shape moral experience, nature, history, and culture are always intertwined. And insofar as humans are biological as well as historical and cultural creatures, human moral experience is rooted in nature.

And second, through innovations in biotechnology, the historic human aim to control the natural environment has crossed a morally significant threshold. As Jürgen Habermas puts this, "more and more of what we are 'by nature' is coming within the reach of biotechnological intervention . . . [and this] is but another manifestation of our tendency to extend continuously the range of what we can control within our natural environment."[4] In other words, the technological extension

---

3. German sociologist Ulrich Beck articulates this moral gap and the challenges of navigating it by specifying the unique character of "risk" in our time. He writes, "Unlike the risks of early industrial society, contemporary nuclear, chemical, ecological, and biological threats are (1) not limitable, either socially or temporally; (2) not accountable according to the prevailing rules of causality, guilt, and liability; and (3) neither compensable nor insurable." For these reasons, he continues, "the regulating system for the 'rational' control of industrial devastation is about as effective as a bicycle brake on a jetliner." See Beck, "Politics in Risk Society," in *Ecological Enlightenment*, 2.

4. Habermas, *The Future of Human Nature*, 23.

of human power into areas such as genetic engineering dissolves the historically entrenched categorical boundary between the surrounding natural environment and the nature of human beings. In an unprecedented way, humanity hereby becomes not only the subject of the power to manipulate nature, but also the object of this power.[5] Understanding this shift is crucial to understanding more fully our human participation within and responsibility for the environmental crisis.

Sociologist Anthony Giddens articulates well this aspect of the environmental crisis in his interpretation of the difference between "external" and "manufactured" risk. By "external risk" he refers to the kind of risk experienced as coming from the outside, from the objective conditions of the natural world and its capacity both to sustain and do harm to humans. For most of modern human history, Giddens writes, this "external" concept of risk fueled scientific and technological development, which were aimed at harnessing and controlling the powers of nature in order to make human life more secure. But, ironically, advances against the pressures of external risk have caused a new kind of risk to emerge, "manufactured risk." Manufactured risk consists of the accumulated threat of technological innovation that now structures human social life. "At a certain point," he writes, "we started worrying less about what nature can do to us, and more about what we have done to nature. This marks the transition from the predominance of external risk to that of manufactured risk."[6]

Put in a different way, in classic Marxist terms, alterations of the objective material conditions of life always impact both the limits and possibilities of human life. As humans articulate the objective world, they articulate their own subjectivity; technologies fashion selves as they fashion the world.[7] Though this dynamic has always been opera-

---

5. Human power's penetration into human biology leads, in Habermas's terminology, to the "dedifferentiation" of the objective natural world and the natural conditions of our human subjectivity. See *The Future of Human Nature*. While we have always been a part of nature, we have historically lived as if this were not the case. Our capacities to alter the natural world have been relatively limited, our impact relatively short term, our biological constitution insulated from fundamental alteration.

6. Giddens, *Runaway World*, 26.

7. Of this double articulation of technological power, its reciprocal influence on the world and upon us, philosopher Langdon Winner writes, " . . . technologies are not merely aids to human activity, but also powerful forces acting to reshape that activity and its meaning." See Winner, *The Whale and the Reactor*, 6.

tive, a time of radical power brings it into high relief and makes moral choices more urgent. The "manufactured risk" of which Giddens speaks is the effect of radical human power and factors into the unique moral dilemmas of the environmental crisis.

What comes to view in light of the insights of Giddens and others is a picture of the environmental crisis with multiple overlapping dimensions, each of which presses its own kind of moral and political choices. It is simultaneously global and local, impacting the whole of the biosphere generally but also specific regions in particular ways. It introduces new social problematics, impacting the possibilities, limits, and ethos of human social interaction as well as human interaction with the natural world. The environmental crisis foments new political questions as well, concerning how societies ought to be organized around a conception of the common good that includes the good of the natural order in which human societies are embedded.

New economic considerations are generated too. Nature is not an inexhaustible resource and factoring it out as an externality falsifies the actual calculus of production and consumption. And there are of course technical and scientific dimensions to the environmental crisis. New technologies and increasing scientific knowledge of our natural world have led to many good things. But with respect to many of the ecological and biological challenges we face, there is constant debate about whether or not these challenges can be met by yet newer technologies and greater scientific understanding.

These complex, manifold consequences of the crisis of human power are *our* human moral burden in several senses. They are ours because they are of human own making. We have manufactured them and so we are causally responsible for them. They are the effect of an increase in scientific knowledge of nature, the development and impact of powerful technologies that allow humans to control and alter the natural world in unprecedented ways, and a failure to choose responsibly how to direct this knowledge and power. Of course not every individual everywhere is equally a cause of environmental problems, but everyone everywhere is impacted in some way by them and every person has some capacity to effect change in the world.[8] All humanity

---

8. Ulrich Beck puts it dramatically: "Until now, all the suffering, all the misery, all the violence inflicted by people on other people recognized the category 'the Other'—workers, Jews, blacks, asylum seekers, dissidents, and so on—and those appar-

is included in the object of moral consideration that the expansion of power compels us to face—the future of all life.

The moral burden of human power does not belong to a specific place, region, tribe, or nation but to all humans everywhere. In this sense, there are no "outsiders" or "marginalized," for all humans belong within the orbit of life's endangerment.[9] The future of life is now an option that must be chosen. And yet this choice is a distinctly human task, for it is a moral choice and only humans possess the moral and cognitive capacities to make it. So in addition to being causally responsible, humanity is morally responsible for the environmental crisis as well.

This radical extension of human power significantly differentiates the "moral space" of this time from other historical periods.[10] We have crossed a moral threshold from the inevitable fact of limited human incursion into nature to the possibility of nature's present and future total alteration. Never before has the future of human life on the planet been an object of moral consideration. That there would be a human future has always until now been a secure assumption. Never before has the future of the entirety of the biosphere been an object of moral concern. That there would be a future for *bios*—for all of life—has until now been a shared confidence, not a matter of choice. Understood in these ways, the environmental crisis demands strenuous and concerted ethical thinking leading to responsible moral participation in the natural world.

---

ently unaffected were safely outside this category. *The advent of nuclear and chemical contamination* [and, he implies, ecological devastation] *has let us experience the "end of the Other," the end of all our carefully cultivated opportunities for distancing ourselves."* See "Survival Issues, Social Structure, and Ecological Enlightenment," in *Ecological Enlightenment*, 27 (italics original).

9. This is of course not to say that ecological degradation is either caused or experienced by all people in the same way. This is emphatically not the case. I simply mean to underline the point that the vulnerability of the natural environment, habitat for the human species among many others, is a broadly human problem.

10. I use the concept "moral space" here in the sense that Charles Taylor uses it in *Sources of the Self*, as an "orientation" that shapes questions about "what is good and bad, what is worth doing and what now, what has meaning and importance . . . and what is trivial and secondary" (28). Part of the usefulness of the concept, as I intend it, is that the spatial metaphor induces moral thinking to move beyond strictly human historical concerns toward a fuller appreciation of our environments, natural and social. See Taylor, *Sources of the Self*, 25–52.

Moral Anthropology and Environmental Ethics

Over the past decades, environmental thinkers and activists have successfully brought the environmental crisis to the attention of many. And yet as German sociologist Ulrich Beck notes, while the precarious state of the natural environment is "accepted in principle, there is still no action, or at best only cosmetic action, taken."[11] There are many reasons for this gulf between perception and action.

For one, the scale of environmental problems is overwhelming and poses a challenge regarding where and how to implement policy. Two other reasons for our moral impotency, related to scale, are causality and accountability. Environmental causality is inherently ambiguous, a puzzle to trace, making it difficult to isolate accountability. Environmental impacts drift, ignoring geopolitical boundaries, crossing public and private sectors. But these sources of inaction are compounded by the internal fragmentation of environmental ethical discourse.

While many environmental thinkers share an ambition to theorize a proper moral relationship between humans and nature, the consensus requisite to effectual response to our environmental challenges is impeded by the conflicting visions among the variety of theories proffered. In short, though environmental thinking has been strenuous, it has not been concerted. Great differences fragment the field. The broader culture's moral uncertainty and its profuse debating are reflected within environmentalism. Conflict rages to the point of paralyzing constructive practical agendas. The environmental philosopher Bryan Norton, for example, has characterized the recent scene of environmental discourse as a "babble of voices," as a "cacophony" of worldviews hampering the process of developing and implementing constructive environmental policies.[12]

Isolating a common concern within this cacophony is an important strategy in the effort to clarify it. I hold that one shared concern in environmental ethics revolves around descriptive and normative accounts of the relation between our human moral nature and the broader natural world, what I will commence to refer to as "moral anthropologies." I would like to stress the *relational* character of what I intend here. With the term "moral anthropology" I do not refer strictly

---

11. Beck, "Politics in Risk Society," *Ecological Enlightenment*, 5.
12. Norton, *Toward Unity Among Environmentalists*, ix.

to an account of human moral nature, or to human moral capacities, though these are features of a moral anthropology. By "moral anthropology" I intend to draw attention to the *relation* between humans and the broader natural world, a relational picture within which description and prescription of this relation circulate and mutually influence one another. That is, a moral anthropological method for environmental ethics describes the human relation to the natural world and informs prescriptions for the way humans ought to relate, for example, to other natural beings, to the evolutionary history of life, and to the broader biophysical environment.

While I take moral anthropology to be a common concern within environmental ethics, even if it is not always explicit, it is manifest as a continuum of positions. The descriptions at either end of this continuum are seemingly irreconcilable. One end of the continuum emphasizes relational continuity between humans and nature, the other discontinuity; while one side understands humanity within nature, the other emphasizes human autonomy and difference. The prescriptions informed by this descriptive continuum also seem to be reversed. Where one end of the continuum can lead to a privileging of the value of natural systems over individuals and species, human and otherwise, the other tends to privilege human value over nature. While one pole affirms humanity's dependence on the natural, the other generally downplays this. Though the influence of a moral anthropology on the broader ethical system containing it is not always acknowledged, it is of profound normative consequence whether an ethical theory emphasizes a description of human continuity or discontinuity with nature.[13]

By "moral anthropology," then, I refer to the reciprocal dynamic between moral imperatives for and descriptions of the human relation to nature. Moral anthropology is an object within the field of ethical theory, encompassing descriptions of the human-nature relation and the prescriptive limits and possibilities those descriptions create for

---

13. Extreme positions on discontinuity, for example, tend to describe nature as radically other, characterizing nature homogenously as nonhuman. This description leads to some version of an instrumentalist axiology in which nature's value is based on its availability for human use. On the other hand, positions on continuity reverse these emphases by downplaying human distinctiveness through a naturalization of human moral capacities. At the extreme end of this tendency, human interests and values have no claim to priority over the natural world. Nature is intrinsically valuable, a center of value itself.

what humans ought to be and do in relation to the natural environment. How a moral anthropology is articulated—what is emphasized and what may be neglected, whether it features an understanding of human morality as continuous with the rest of natural life or as marking a strict divide—impacts theories of value and obligation and the feasibility and limitations of policy. Given this, moral anthropologies are bright stars in the various constellations of ethical theory and can serve as a helpful interpretive, comparative, and critical focus of study.[14]

While I hold that the interpretive significance of moral anthropology stands for ethical theory in general, this significance is especially pronounced and of enormous practical relevance in the field of environmental ethics. Against the sensibility of many environmental thinkers, I contend that environmental ethics should move towards rather than away from an anthropological focus.

Some environmental thinkers critique any focus on the human as pejoratively anthropocentric. But the automatic assumption that any concern with the human is out of place in environmental thinking radically narrows the field of environmental ethics and minimizes the moral challenges of the environmental crisis. Along with theologian Peter Scott, I understand environmental concern as a human concern, it "is not directed to some abstraction, called Nature. Instead it is directed towards the quality and character of habitation, including the habitation of humanity."[15] The fundamental concern of environmental thinking, in my judgment, is how morally to live the good and right life in relation to the natural environments that we simultaneously depend upon and threaten.

Humans, then, are central to the project of environmental thought. Because the human is at the center of the environmental crisis, insofar as this crisis is directly related to the extended scale of human power and the rise of moral uncertainty, the human needs to be at the center of an ethical response.

---

14. Along with religious ethicist Anna Peterson, I affirm that "any idea of human nature has ethical implications and that all ethical systems rest upon certain ideas of human nature." However, as Peterson notes, this does not mean that anthropology alone is sufficient to the understanding of ethical theories. It suggests instead that as a necessary dimension of ethical theory, granting a measure of critical attention to moral anthropology can advance the understanding of ethical systems. Peterson, *Being Human*, 3.

15. Scott, *Political Theology*, 3.

In light of the moral pressures of the present ethos, environmental ethical thinkers, both philosophers and theologians, urgently need to reexamine some old questions. Questions concerning the character, meaning, and responsibilities of human power and moral agency, questions integral to any account of moral anthropology, need to be faced. Such questions stretch deeply back into human history. But the reach and magnitude of human power has changed and thus makes these questions more urgent.

Like all living beings, humans influence the world around them. Humans have always had a greater capacity to impact the world than other species, but the capacities we now possess outstrip all previous historical periods. If the ethos of this age is defined at least in large measure by the human capacity radically to change the natural environment, then a concern with moral anthropology seems to have an important place in contemporary environmental ethical theory and practice.

A moral anthropological method for environmental ethics is also requisite today because the traditionally assumed divide between humans and the rest of nature has been severely challenged in the last century and a half by work in the biological and ecological sciences. The human relation to nature has always been and is continually being redescribed. Certainly humans are social beings, inhabiting cultures, building and critiquing institutions, living in response to and making history. But humans are also beings of and in nature. The scale of human efficacy and the choices humans make and fail to make impact the present and future of humanity, the present and future of all life, and require critical ethical consideration in light of the changing understandings of the human relation to the natural world.

Since the publication of Darwin's *Origin of Species*, many traditional understandings of the human relation to nature have been called into question. The assumption of radical human discontinuity with the natural world has largely been replaced by a concern with how to understand humans in relation to other forms of life. Given current scientific knowledge, it is difficult if not irresponsible to maintain the position of a thinker even as great as Rabbi Abraham Heschel. For Heschel, "We can attain adequate understanding of man only if we think of man in human terms, *more humano*, and abstain from employing categories developed in the investigation of lower forms of life."[16] To claim to the contrary that an adequate moral anthropology requires employing the insights of the biological sciences need not be viewed as a degradation of humanity.

Why should humans be ashamed of kinship with other animals, or our relation to the supposedly "lower" forms of life? Is it possible to understand either "being human" or "human being" in isolation from the human habitat?[17] Heschel's claim betrays a morally troubling, historically deep descriptive and axiological dualism between humans and the rest of nature, as well perhaps as a warranted critique of overzealous scientific reductionism.

Appropriate in response to Heschel is the claim of philosopher Mary Midgley that the proper question to ask nowadays is not what distinguishes humans *from* other animals but what distinguishes us *among* them.[18] It is impossible to deny that humans are animals, in addition to everything else we may be. Whatever we focus on as distinctive about our nature, whether reason, language, or our moral capacities, needs to be understood "not against the laws of our nature, but according to them."[19] While Darwin's work initiated a revolution in our understanding of human nature, debate continues to rage about its philosophical, ethical, and religious implications. That we are natural beings is granted. But the question of our precise location on the continuum between continuity and discontinuity with nature and the normative significance of this question remain open to debate. What is the role of genes in human nature and behavior? What is the role of culture in how we identify ourselves as agents, in what we value, in the practical moral character of our lives? To refer to the biologist E.O. Wilson's evocative metaphor, is there a genetic "leash" on culture, and if so, what length?[20] If it is crucial to consider human distinctiveness in accord with the

---

16. Heschel, *Who is Man?* 3.

17. With this question I refer to Heschel's distinction between *human being* and *being human*. *Human being* for Heschel is a category within the class of animals, but *being human* for him is irreducible to this status. The question I raise here, contra Heschel, is whether whatever it is that one takes to be distinctive about *being human* can be understood or have meaning in isolation from the animal nature of *human being*.

18. See Midgley, *Beast and Man* and *The Ethical Primate*.

19. Midgley, *The Ethical Primate*, 24.

20. See Wilson, *On Human Nature*, 167.

natural laws of nature, as Midgley suggests, what measure of control and degree of influence do these laws exert?

In thinking through the questions of human power and moral uncertainty through a moral anthropological framework, it is important to try to balance human continuity *and* discontinuity with nature. Against the tendency to emphasize one pole of this continuum over the other, an effort must be made to hold the insights of both views together. For even if humans are a particular mammalian species of animal in continuity with other evolved forms of life, we possess important species-typical characteristics that must be considered if we are really to know ourselves. Evolutionary continuity does not warrant but rather challenges biotic sameness. To disregard human difference is to disregard the insights of biological theory about the typical characteristics of our species and is just as biologically misinformed as a stress on radical discontinuity.

But such disregard for human distinctiveness is much more than a descriptive problem. It is most of all a deeply moral one. Disregarding human difference naïvely neglects the unique and empirically undeniable capacities of the human species to alter the world more radically than any other animal. While every animal of course exerts an environmental impact, only humans face the moral choice of whether to change totally, even to destroy the whole planet. The pressure of this moral choice suggests that a central ethical task of this time is to theorize the human relationship with the larger natural environment without eclipsing the enormous practical relevance of human distinctiveness. What is needed, against the tendencies to prioritize either continuity or discontinuity, is a moral anthropology that holds together commonality and distinctiveness. As will soon be explained, the work of Hans Jonas and James Gustafson provide suggestive resources for a moral anthropology that dialectically affirms continuity and discontinuity, both kinship with and differences from other forms of life.

For this reason and others, responding to the environmental crisis entails entering the "tangled bank" of scientific, philosophical, and theological discourses about the human moral relation to the natural world. This interdisciplinarity is a reflection of the fact that knowing what humans are and who we can be as moral beings is of necessity a multiperspectival task, relating to the various dimensions of human experience and to the various natural and cultural tributaries of the human moral condition. Humans are not merely cultural, social, biological, rational, language-using or religious creatures. Humans are all of these things and more, inhabiting natural and cultural environments that we shape but that also shape us.

Two Trajectories of the Argument

My argument in this book unfolds along the lines of two trajectories. One of these entails a comparative critique of the environmental ethical projects of Jonas and Gustafson with a central focus on their moral anthropologies. Attending to these thinkers' moral anthropologies serves as an interpretive key to their broader environmental ethical theories and brings into relief their insights and liabilities. Though generally overlooked by the field, Jonas and Gustafson have much to contribute to the sophistication of environmental ethical discourse. Each of them develops their theories with a strong sense of the significance of moral anthropology. And very importantly, they are both keen to the problems of tilting too far toward either human continuity or discontinuity with nature. In addition, though each uses the biological and environmental sciences differently, each seeks to present his ethical theory as appropriately informed by the natural sciences.

Sensitive to the magnitude of contemporary human power and human moral responsibility to the natural environment, Jonas and Gustafson both attempt to hold together human biological continuity with nature and concern for the moral distinctiveness of the human within nature. They undertake their projects in relation to the "tangled bank" of human biology, religiosity, and morality. In this process they aim to do justice to the complexity of the human relation to nature in our technological culture and in a time of environmental crisis. In these various ways, then, Jonas and Gustafson are thinkers whose works engage some of the most significant tasks of environmental ethics.

However, my burden is not simply to interpret what Jonas and Gustafson themselves have already said so well, though since they have been generally unappreciated, this in its own right is a valuable exercise. While I do grant significant attention to their theories, my ultimate purpose, building upon their insights, is to recommend an ecotheological ethics of responsible participation as a promising framework for future theological work in environmental ethics.

My aim is not to present or defend a wholly new ethics. Rather, I intend to develop the insights of Jonas and Gustafson in order to suggest an understanding of the human moral relation to nature that is biologically grounded and theologically and morally viable. I believe that part of the promise of this ecotheological ethics of responsible participation is in the way that it can account for the moral and theological significance of our participation within and our responsibilities for natural processes, and in doing so challenges the false opposition between anthropocentrism and nonanthropocentrism in environmental ethics. But before proceeding, I want to conclude this introduction by delineating the theological dimension of my project through a summary of the differences and complementarities between Jonas and Gustafson's theological orientations.

According to Jonas's own claims, theology plays only a peripheral role in his work. He does not dismiss theology outright but argues instead that while his philosophical vision is not incompatible with certain theological convictions, it does not require theological backing. As he puts it, theology is "a luxury of reason" unnecessary to the integrity of his theory. Furthermore, he claims that ethics generally ought not to turn to theology for an account of moral motivation. His point, as I will discuss in detail in later chapters, is that being's value is intuitively obvious, human experience testifies to it, and the intuition of this value alone should be sufficient to motivate and bind responsible human action toward the natural world. On his own account, then, theology is treated as an "adjunct" of morality.[21] This placement of theology in Jonas's work reflects the centrality he grants to what he calls "the idea of humanity" in the normative dimension of his ethics.

Though Jonas's stated ethical methodology suggests this marginal role for theology, this is somewhat betrayed by what I interpret as theology's constructive function in some of his writings. Indeed, Jonas's profound concern for the vulnerable futures of the natural world and human life, I will suggest, lead him to an appreciation for the significance of the theoethical imagination. I affirm Jonas's views here and his naturalistic defense of them and think that his insights can constructively supplement Gustafson's.

---

21. I borrow this apt term from Sheila Davaney's characterization of Kant's view of theology in his ethics. See *Pragmatic Historicism*, 6.

In contrast to Jonas, Gustafson's ethics is very self-consciously theological. He argues for the hermeneutic and ethical centrality of what I will specify later as a critical religious naturalism. The patterns and processes of nature so central in Gustafson's theory serve as empirical indications of the divine ordering of the world, and thus discerning their moral valence is a crucial task of the religious life. Gustafson's controlling methodological criterion is that theological claims cannot be significantly incongruent with the claims of science. This leads him to reject granting theological claims a priori status. In place of this, he deploys a methodological strategy that entails critical movement back and forth between theological tradition, human experience, and scientific knowledge. Gustafson's views on the unity of knowledge, the common structures of human experience, and the critical congruity of theological and scientific discourses are strongly featured in this strategy.

The overall character of Gustafson's theology can thus well be described in terms of the revisionist trajectory in some recent theology. Unlike Jonas, Gustafson treats moral questions within a decidedly theological context. And yet against a strong revelationalism or biblicism, he does not treat the biblical narrative or the deep grammar of the Christian tradition as theologically determinative for the moral life. While confessing his indebtedness to the Reformed theological tradition, his theology is not parochially confessional. Instead, it is more in keeping with a lineage associated with theologians Paul Tillich and David Tracy in which theological traditions remain open and revisable in light of contemporary experience and knowledge.

Given this lineage, Gustafson's theological task entails walking a fine line. He aims to demonstrate the public significance of his vision while not abandoning its distinctive theological character. Put differently, he aspires to communicate the relevance of his theological ethics to broad public moral concerns, such as the environmental crisis, without distancing himself too far from the theological tradition he claims to represent. I strongly affirm this aspiration and will explore in later chapters the degree to which it can be said that Gustafson attains it.

Despite these very different theological orientations, there are ways in which the theological aspects of Jonas's and Gustafson's projects can be interpreted as complementary. For one thing, both operate with deeply naturalistic methodological commitments. Jonas's naturalism is shaped largely by a phenomenological existential tradition in philosophy, and yet remains open to theological insights. While Gustafson's naturalism is shaped by a revised Reformed theological tradition, one of the strengths of his work is the degree to which it invites interaction with non-religious philosophical interlocutors as well as with other religious visions.

And second, the theological insights of both Jonas and Gustafson, different as they are, are strongly influenced by a common concern for the future of nature and the human moral place within it. As I will develop more fully later, I take the constructive role of theology in Jonas's work to be rooted in his sense of the moral urgency of the environmental crisis. In light of this, and somewhat against his own claims, his imaginative theological work becomes a prominent task of environmental moral responsibility. Gustafson's theological vision is similarly influenced by a vital moral concern for the natural world. The theocentric vision that arises in part out of this concern becomes a lever for Gustafson's revisions to the way humans think and speak about God and the human good.

Along with both thinkers, this book affirms concern for the plight of the natural world as a crucial point of departure for theological ethics in our time. Through comparative critique of Jonas's and Gustafson's projects, the chapters to follow point toward an ecotheological ethics of responsible participation as a fertile option for this work, one that is as deeply naturalistic and as deeply theological as the moral pressures of our time demand.

PART ONE

Hermeneutical Dimensions

1

Human Power and Existential Biology

Introduction

SHORTLY BEFORE HIS DEATH IN 1993, HANS JONAS PRESENTED A TALK on the occasion of receiving the Premio Nonio for *The Imperative of Responsibility*, the award given to him as author of the best book translated into Italian that year. The award's sponsor had asked him to reflect on the topic of racism. As a German Jew who abandoned an academic career in his homeland in order to volunteer with the British Army in its resistance to Nazi Germany, whose mother was murdered at Auschwitz, Jonas was uniquely qualified to discourse on this topic. He began by sharing a brief and moving narrative recalling an encounter with two Jewish women in Italy at the end of the war, in the very city of Udine where he was now receiving his award. In concluding this narrative, however, he declared the problem of racism to be "anachronistic, irrelevant, almost farcical before the all-encompassing challenge which an endangered global environment flings in the face of all mankind."[1] This challenge of an endangered planet, argued Jonas, transvalues the issue of unjust social exclusion upon which racism pivots by forcing upon all humans a radical sense of solidarity. Under the pressure of global ecological catastrophe, humanity is one and not many. Of ecological holocaust, each of us is simultaneously victim and victimizer, each and all of us together the subject and object of a crisis that relativizes all other crises. Jonas prophetically summarized his appraisal of our endangerment in his final public comments with these words:

> It was once religion which told us that we are all sinners, because of original sin. It is now the ecology of our planet which

---

1. Jonas, *Mortality and Morality*, 201.

pronounces us all to be sinners because of the excessive exploits of human inventiveness. It was once religion which threatened us with a last judgment at the end of days. It is now our tortured planet which predicts the arrival of such a day without any heavenly intervention. The latest revelation—from no Mount Sinai, from no Mount of the Sermon, from no Bo (tree of Buddha)—is *the outcry of mute things* themselves that we must heed by curbing our powers over creation, lest we perish together on a wasteland of what was creation.²

This statement not only expresses Jonas's view of the gravity of our ecological crisis, but also powerfully reflects three key stages of his personal and intellectual history.³ The juxtaposition of otherworldly religious worldviews with the crush of this-worldly planetary realities reflects his early scholarly interest in the world-denying themes of Gnosticism; his historical thesis that Gnosticism's metaphysical dualism continues to exert a crippling influence on Western philosophy; and his eventual movement from the study of Gnosticism to the philosophy of biology and ethics. The rhetorical displacement of religion's threatening supernatural judgment by the natural threats of ecological and human vulnerability reflects his insight that the force of our moral obligations stem from the ground of being's intrinsic perishability. And the assertion that the "cry of mute things" contains the latest revelation parallels his core ethical claim that the binding nature of and motivation to moral responsibility hinge on our perception of biotic vulnerability—on regaining the capacity to hear the true eloquence of living nature itself—rather than on theological backing. In short, this quote registers the unity and originality of the three-staged sequence of Jonas's

---

2. Ibid., 201–2 (italics mine).

3. Lawrence Vogel, in his thoughtful introductory essay to *Mortality and Morality*, similarly interprets three major phases of Jonas's thought. However, Vogel's treatment isolates existential, metaphysical, and theological stages. These stages are united, on Vogel's account, by Jonas's concern to challenge the nihilism of his mentor, Martin Heidegger. The difference between the stages as I present them and Vogel's interpretation turns on my thesis that, however important Heidegger may be to Jonas's development as a thinker, the hermeneutical key of moral anthropology is crucial to an understanding of the originality and coherence of Jonas's wide-ranging thought. As the remainder of this chapter will elucidate, Jonas is intensely concerned with nihilism, for personal as well as scholarly reasons. But his response to this concern is to develop an ethics of responsibility, rooted in what I will later examine as the moral juncture between his moral anthropology and philosophy of nature.

thinking: his early existential analysis of Gnosticism, his work on the philosophy of natural life, and his movement toward ethics.[4] Uniting the seemingly disparate stages in this sequence, however, is an underlying concern with the nature and burdens of human power and choice.

In this chapter, my aim is to clarify and to link Jonas's moral motivation and methodology, what I take to be the hermeneutical context of his work. The first part of the chapter attempts to isolate Jonas's moral motivation through an interpretation of the historical roots of the contemporary ethos. The second analyzes the method he develops in his constructive response, what he calls an "existential interpretation of biological facts."[5] Through these two exegetical tasks, this chapter aims to show the mutual influence between Jonas's conception of the fundamental moral problem of the time and his methodology, pointing to the way this lends insight into the strengths and weaknesses of the

---

4. Gereon Wolters also affirms the unity of these stages in Jonas's thinking, and the significant role of what I refer to as his moral anthropology. Wolters writes,

> Contrary to first appearances, these three topics stand in close relationship to each other. Jonas's philosophical biology is an attempt to overcome the dualism, i.e., the alienation between man and world, which characterizes both Gnostic thinking and the Heideggerian existentialist approach that Jonas has applied in its interpretation. This dualism leads both approaches to despise or, at least, to neglect nature . . . . The attempt to overcome this dualism between man and world is at the very heart of Jonas's philosophy. It consists of two steps: finding a place for human nature, both ontologically and anthropologically, in the world; and second, developing an ethics with this ontological basis as its foundation.

See Wolters, "Hans Jonas's Philosophical Biology," 87–88, first presented at the Arendt/Schürmann Memorial Symposium on the Philosophy of Hans Jonas at New School University, Fall 2000.

5. The meaning of this methodological terminology will be more fully addressed further along, but for now it can be said that in describing his approach to biological facts as an "existential" one, Jonas intends a phenomenological reworking and extension of the ideas of existentialism, especially his interpretation of Heidegger's existentialism, by way of the insights of biological theory. The thesis of life's evolutionary continuity specifically leads this reworking, for whatever can be said is characteristic of human inwardness must, on the thesis of continuity, be present in some way in less complex forms of life. Thus, by an "existential approach to biological fact" Jonas aims to appropriate the significance of a phenomenological emphasis on the inward testimony of human consciousness for the insights it can yield into the purposive character of all of natural life. In essence, rather than taking freedom and the senses of peril and dread as characteristic of human existence alone, Jonas identifies traces freedom through all of life and finds life as such to exist as a precarious negotiation of the crisis of being and not-being.

other aspects of his thought, and establishing a basis for comparison with the hermeneutical dimensions of Gustafson's.

The Ambiguity of Power

To grasp fully Jonas's moral diagnosis of our ethos, it is important to begin where he began, with a summary of his early historiography of gnostic dualism. This will naturally lead into a discussion of his critique of the philosophy of his early teacher Heidegger, since Jonas's interpretation of Gnosticism was deeply influenced by the existentialist sensibility he inherited from Heidegger. Together, these tasks will bring to surface the problem that drives his work, the moral ambiguity of contemporary human power.

Gnostic Dualism

Foreshadowing the concern of his later work, Jonas writes at the conclusion of his 1952 essay "Gnosticism, Nihilism, and Existentialism," added as an epilogue to the second edition of *The Gnostic Religion*, that, "the disruption between man and total reality is at the bottom of nihilism."[6] Supporting the assumption that morality redounds to power is, Jonas here claims, a dualism between humanity and the natural world. In this same essay Jonas acknowledges that the modern philosophical optics of existentialism, and most specifically the work of Heidegger, allowed him to make a unique contribution to the study of Gnosticism, and, reciprocally, that his study of Gnosticism led him to greater insight into the contemporary philosophical scene. He writes, "The extended discourse with ancient nihilism proved—to me at least—a help in discerning and placing the meaning of modern nihilism: just as the latter had initially equipped me for spotting its obscure cousin in the past."[7] This claim marks the creative dialectic in Jonas's thought between historical retrieval and his constructive philosophy.

The content of Jonas's work on Gnosticism is at best only indirectly focused on the problem of natural life that occupies him through the

---

6. Jonas, *Gnostic Religion*, 340.

7. Ibid., 320. Jonas's existentialist reading of Gnosticism was influenced and encouraged by his great teacher and mentor, Rudolf Bultman, known for his existentialist demythologization of the Christian New Testament, along with, as already indicated, his other great teacher, Martin Heidegger.

second and third stages of his career. But his decidedly philosophical and existential treatment of Gnosticism, in contrast to a more strictly historical approach, contains the seed eventually flowering into his later concerns. It allowed him to discern the metaphysical similarities underlying the ancient and modern philosophical situations and to set the groundwork for a practical ethical approach to these issues. Jonas's approach to Gnosticism was motivated by his perception that though gnostic themes have been neglected as a philosophical topic of study, these themes still significantly shape the modern philosophical ethos and our contemporary lived reality and therefore warrant careful examination.

Especially resonant with Jonas was the gnostic theme of radical metaphysical dualism. The basis for this resonance was his thesis that the existentialist view of the human as basically alienated from the world is not an explication of the fundamental nature of humans in the world but a response to particular historical contingencies. To the degree that Jonas can show that the historical contingencies underlying modern existentialism parallel the situation of the ancient gnostic movements, then a juxtaposition of existentialism and Gnosticism offers the possibility of the mutual illumination of each. The history existentialism and Gnosticism have in common, Jonas will argue, is a history imbued with a dualistic metaphysics than an ethics in an age of radical power must overcome.

In the essay cited above from *The Gnostic Religion*, Jonas locates the origins of modern existentialism in Pascal, who with eloquent anxiety described the loneliness of humanity in relation to a world described by modern cosmology as utterly indifferent to human purposes. According to Pascal, influenced as he was by the Cartesian dichotomy of *res extensa* and *res cogitans*, existential dread results from recognizing that the human is an anomaly within the universe. Humans, the only thinking substances in the universe, are by their very essence estranged from the world. This estrangement, rooted in modern science and Cartesian dualism, becomes basic to the existentialist interpretation of the human condition. It is the metaphysics of an indifferent universe posited by modern science that leads to a picture of the human condition emphasizing "homelessness, forlornness, and dread."[8]

8. Ibid., 323.

The similarity underlying this modern account of the human condition and ancient Gnosticism is based on some common features in their philosophies of nature. For both existentialism and Gnosticism, there is no basis for kinship between humanity and the universe. Though Jonas recognizes that Gnosticism is not monolith, but is rather a collection of movements and traditions, and that its origins are geographically and religiously dispersed, gnostic variety was united by several key themes.

The most significant of these themes was the radical metaphysical dualism mentioned above. Gnostic dualism had three basic dimensions according to Jonas: it described the human relation to the world, the relation between the world and the divine, and the relation between the divine and humans. On Jonas's interpretation, dualism was psychologically manifest as the feeling of dread resulting from a sense of the "absolute rift" between humanity and the world. Theologically, gnostic dualism shaped a doctrine of God as wholly and totally other, unknowable through "worldly analogies."[9] Cosmologically, the correlate of dualism was the view that the world was the creation not of God but of an inferior and evil demiurge. According to this account, the world is fundamentally malignant. As the creation of the demiurge, not of God, the world is a product of ignorance rather than knowledge, the expression of power and force rather than of reason, the domain of darkness rather than light.[10]

Anthropologically, the situation of humans within this ethos of cosmic antagonism was that of homelessness and dread, and was paralleled by a physiological homelessness as well. The body as material was viewed as foreign to the spiritual essence of human nature. Gnostic anthropology treats both the body and soul as material and mundane, encasing the spirit or pneuma, understood as a portion of divine substance, within the world's hostile physicality. Salvation from the hostility of the world and from one's own body and soul depended on *gnosis*. The aim of *gnosis* was not to overcome alienation, but rather to overcome the world. Alienation, as will be described below, was not

---

9. Ibid., 327.

10. In Jonas's judgment, this world was yet a cosmos, though in a sense that was utterly alien to the traditional Greek understanding of cosmos. Replacing the earlier Greek emphasis on the logos of the universe, Gnosticism understood the tyrannical power of the world as *heimarmene*, universal fate. The gnostic view of fate was of a hostile force, governed by the evil demiurge and his minions.

something to be transcended but awakened to. Salvation depended on the Savior's forceful intrusion into the world, the dissemination of his knowledge, and the use of that knowledge to combat the alien power of the world.[11]

As the earlier reference to Pascal indicated, the metaphysics underlying modern existentialism is rooted in a Cartesian and Newtonian scientific view of nature that, similarly to Gnosticism, posits an ontological split between mind and materiality. Though the ancient gnostic view of nature and the physical is pronouncedly negative, Jonas argues that it is not as morally dangerous as the modern. I will now turn to this through an examination of Jonas's critique of one of his most influential teachers, Martin Heidegger.

Heidegger's Nihilism

As already indicated, Jonas's excursion into the study of Gnosticism was based in part on his view that it shared significant similarities with modern existentialism, and that a study of these similarities would shed new light on the ancient movement. These similarities, according to Lawrence Vogel, come down to two: "(1) the denial that the cosmos is ordered for the good and (2) a belief in the transcendence of the acosmic self."[12] In addition to these basic similarities, Jonas acknowledges a key difference between the two movements. On the gnostic account, shaped by a malign acosmism, the human exists in a world that is essentially hostile. But on the existentialist account, shaped by the Cartesian dichotomy, the world is wholly indifferent to humans. As pure extension, mindless and without purpose, nature provides no ethical measure for human action, neither a negative nor a positive standard.

This difference between nature's hostility and its indifference marks the unique danger of the modern situation for Jonas. Though the gnostics interpreted nature as antagonistic, nature yet exerted and possessed a moral dimension. For the gnostics of course the moral dimension of nature was negative, but the most severe moral problem, argues Jonas, lies in a view of nature as indifferent or morally neutral. He

---

11. In ch. 5 and in my conclusion, I will argue that there is an ironic and troubling parallel between the Gnostic account of salvation and Jonas's own political prescriptions.

12. Vogel, "Hans Jonas's Diagnosis of Nihilism," 57.

writes, "This makes modern nihilism infinitely more radical and more desperate than gnostic nihilism ever could be for all its panic terror of the world and its defiant contempt of its laws. That nature does not care, one way or the other, is the true abyss."[13] The true abyss stems from the radical cleavage between being and value, which results in an absence of ontological support or sanction for any human purposes. Without metaphysical backing, without objective grounding in the nature of things, meanings, values, and norms are no longer envisioned or registered but become projects of the human will to power. Power without direction or limitation holds sway; power is value's only recourse.

Jonas's initial concern with the unique character of the modern moral abyss, stemming in his judgment from existentialism's underlying philosophy of nature, was not related so much to an interest in an ethics of nature but rather to an effort to understand the source of Heidegger's capitulation to Nazism. According to Jonas's interpretation, his teacher's susceptibility to Nazism was a symptom of the ethical vacuum resulting, in part, from his acceptance of the Cartesian dualistic metaphysics. But this is strange considering that Heidegger's project was motivated by an effort precisely to overcome the Cartesian categories operative in Husserl's phenomenology, and the impediment of those categories to an understanding of being.

Jonas's critique of Heidegger turns on his view that Heidegger's deployment of the analytic of *Dasein* in his effort to overcome Cartesian dualism is limited to the existential experience of interhuman relations, and therefore itself reflects the very dualism it aims to move beyond. Thus, according to Jonas, Heidegger's philosophy does not represent a radical turn from Cartesian metaphysics but is instead its culmination. Heidegger had contributed a significant diagnosis of this dualism, and yet, ironically, the expression and practical implications of this diagnosis were simultaneously in collusion with it. At the core of Heidegger's supposed turn toward being there remains, for Jonas, an ontological bifurcation of the human and natural.

Jonas's most challenging critiques of Heidegger are leveled precisely at Heidegger's accounts of Dasein's temporality and the idea of authenticity, the places where he identifies the strongest evidence of his mentor's complicity in a dualistic metaphysics. Heidegger claims that

---

13. Jonas, *Phenomenon of Life*, 339.

when Dasein becomes lost or absorbed within the seemingly recalcitrant distractions of everydayness, it becomes false with respect to itself and exists inauthentically. That is, it fails to be most fully its own, actively and dynamically pursuing its ownmost possibilities. To exist inauthentically is to allow one's own fundamental ontological relation to the meaning of being to become occluded by ontic concerns. Authentic existence depends on awakening to and enacting one's ownmost possibilities; it is the project of gathering one's temporality unto oneself and thereby recognizing the distinctive mode of Dasein's being in the world.[14]

On Jonas's reading, Heidegger is making a highly problematic assumption here—although the purpose of the accounts of authenticity and inauthenticity may be only indirectly related or propaedeutic to Heidegger's larger project, authentic existence cannot be ascribed to any form of life other than human. Jonas's critique of this emerges most forcefully in his reading of Heidegger's *Letter on Humanism*. In the *Letter*, Heidegger makes the case for an alternative reading of the human against the grain of classical anthropologies. According to Heidegger, the human has been dehumanized to the extent that it has been treated within the categories of classical substance metaphysics. Such treatments dehumanize Dasein by neglecting its distinctively temporal mode of being. According to Jonas, the nihilistic implications of Heidegger's philosophy appear precisely here, in the case Heidegger makes for the radical distinctiveness of Dasein against the presumably dehumanizing anthropologies tethered to classical metaphysics. For in doing so, Jonas writes, Heidegger is rejecting "any 'definable' nature of man which would subject his sovereign existence to a predetermined

---

14. This discussion of Heidegger's distinction between authenticity and inauthenticity can and often is interpreted in an existential fashion. But for Heidegger this is not so much an existentialist declaration as a crucial part of the project of analyzing Dasein, undertaken for the purpose of gaining access to the question of fundamental ontology, the meaning of being. The essence of Jonas's critique is that, regardless of whether the issue of authenticity is interpreted as an existential evaluation or not, Heidegger posits a radical cleavage between the lives of human beings and the being of other forms of life. As will be discussed, Jonas does not take issue with the idea that a phenomenology of human life provides an epistemic opening into fundamental ontology. In fact, he does something very similar himself, treating human inwardness as a point of access into the broader world of life. The problem for Jonas is that Heidegger marks the line between human and other life so boldly that the human becomes severed from a normative relation to nature.

essence and thus make him part of an objective order of essences in the totality of nature."[15]

The issue for Jonas is not so much that classical metaphysics provides the only framework for an adequate anthropology, but that Heidegger's image of the human depends on rejecting a normative picture of human nature and of the rootedness of human life in the natural world. The logic of this rejection and its moral problematic derives from the bifurcation of the human and the larger order of natural life that Jonas takes to be at the heart of Heidegger's project, reflecting his judgment that Heidegger's philosophy is a continuation rather than an overcoming of the distortions of metaphysical dualism.

For Jonas, this bifurcation of the human and other modes of life parallels the gnostic separation of matter and spirit, but accrues significantly greater negative moral gravity to the extent that the human is drained of any nature whatsoever. He writes, "In this conception of a trans-essential, freely 'self-projecting' existence I see something comparable to the gnostic concept of the transpsychical negativity of the pneuma," and the danger of this on Jonas's reading is that what "has no nature has no norm."[16] In Jonas's view, then, the nihilistic consequences of his mentor's philosophy result from Dasein's utter disconnection from the order of nature within and outside of human being. Dasein's "essence," to put it paradoxically, is to have none. For Jonas, the existentials of Dasein mean that Dasein is self-constituting, self-originating, trans-essential and that authenticity is the project of free self-projection unconstrained by any norm in nature or historic tradition. Authenticity is not conformity to any given norm or the fulfillment of any prior nature, but is the self's normless expression of its power to invent itself.

Jonas notes not only the moral danger of this view, but a conceptual paradox at its core. For though the concept "Dasein" literally means "there-being" or "being-there," the self-constitutive nature of authentic Dasein implies, for Jonas, that *no there is there*. In spite of including the exstases of past, present, and future in his analysis of Dasein's existentials, Jonas argues that *the present* actually has no independent ontological status in Heidegger's schema. The "present" for Dasein is a cipher,

---

15. Jonas, *Phenomenon of Life*, 228.
16. Ibid.

merely a function of the "breathless dynamism" of past and future.[17] Genuine or authentic ek-sistence is solely a function of the dynamic between past and future, a standing out from the present. When the present is discussed, it is discussed as a " 'situation' . . . wholly defined in terms of the self's *relation* to its 'future' and 'past.'"[18] The "present" is reduced to mere function of the horizons of past and future, leaving Dasein, on Jonas's reading, without a there in which to be.

Jonas's insight into Heidegger's philosophy includes not only the recognition of the nihilistic implications of his anthropology, nor only his perceptive account of the paradox at the heart of it, but, as was earlier suggested, his understanding of the dualistic metaphysics underlying it. In contrast to the presumed, but on Jonas's reading inconsistent "being-there" of Dasein, beings and things other-than-human for Heidegger are either *zuhanden* or *vorhanden*. In contrast to things that are *zuhanden*, useable or available in some way for the work of Dasein's eksistence, things characterized as *vorhanden* are merely extant, irrelevant to Dasein and thus indifferently present. Irrelevant, such things "are stripped and alienated to the mode of mute thing-hood."[19] Such mute reification, on Jonas's reading, is the status Heidegger's philosophy bestows to nature. Nature for Heidegger is categorically "a deficient mode of being."[20] Jonas charges that, "No philosophy has ever been less concerned about nature than Existentialism, for which it has no dignity left . . . ."[21] Nature is morally neutralized. This, for Jonas, is the deep abyss of modern nihilism, leaving the radical capacities of contemporary human power without a grounding normative vision.

Jonas's Moral Motivation: A Summary

Jonas's earliest work identifies a common theme of dualism underlying the ancient gnostic movements and Heidegger's philosophy. For Gnosticism, this dualism is moral, represented as a conflict between the divine power for good and the power for evil of the demiurge. For Heidegger, on Jonas's reading, this dualism is manifest as a dichotomy

17. Ibid., 231.
18. Ibid., 230.
19. Ibid., 231.
20. Ibid.
21. Ibid., 232.

between the value-generating character of human being and the value-neutral character of the larger natural world. This modern view, according to Jonas, is more problematic morally than Gnosticism. Value becomes merely a human projection, objectively unhinged. Only humans have the capacity for authentic existence; nature is neither good nor evil but indifferent and merely extant.

The morally problematic nature of this vision, for Jonas, is that if value is only a function of human power, then there is no objective or external standard for prescribing or proscribing the exercise of power. This is the essence of the problem motivating Jonas's work: the threat of human power's moral ambiguity. In an age of tremendous human powers, if the only source of value is power itself, then everything is threatened. With the enormous human capacities to alter the present and future world, power needs normative direction.

As I have shown in the preceding, an historically entrenched dualism between humanity and the natural world underwrites this moral problem, and for Jonas, this dualism is traceable from the ancient Gnostic vision up to the work of Heidegger. As I will discuss in what follows, Jonas's response to the problematic of human power begins from the position that the dualism upon which it rests can be shown to be internally inconsistent according to the Darwinian thesis of evolutionary continuity. Given evolutionary continuity, Jonas asks, how can an indifferent nature give rise to a form of being that is characteristically self-concerned? And so in what follows I move from the moral motivation backing Jonas's work to an examination of his methodological commitments.

Jonas's development of an "existential biology" levels a strong blow to the dualism he sees underlying both Gnosticism and Heidegger's philosophy and is strategic to his larger aim of developing a moral anthropology and norm of responsibility. He aims at nothing less than a unified philosophy of nature, one that includes the human within the natural, in order to provide scientific and philosophical support for an ethical vision of responsibility toward nature.

Existential Biology

Jonas's methodology turns on the idea that modern scientific materialism contains the seeds of its unraveling in its own greatest triumph,

Darwinian evolutionary theory. On Jonas's interpretation of evolutionary theory, for mind to exist anywhere in nature requires that it must be prefigured in earlier forms of life. Therefore, mind and matter are not antithetical. For Jonas, Darwinian theory essentially implies human continuity with the rest of nature. Whatever human capacities have been traditionally conceived as distinguishing the human *from* the rest of life become, through Darwinism, capacities that characterize the human *among* other forms of life. Put differently, traces of subjectivity must be present in life's most rudimentary forms if one grants the basic thesis of evolutionary continuity.

With this basic idea Jonas aims to recover through an "existential interpretation of biological facts" the inner dimension of bios neglected by anthropocentric existentialism and the physicalist prejudice of modern science. In spite of profoundly different outcomes and motivations, Jonas adapts elements of Heidegger's philosophy in his reinterpretation of nature. He begins, as Heidegger began with Dasein, from a phenomenology of human subjectivity. But in place of discontinuity between human and other forms of life, Jonas interprets a progressive and continuous scale of life's complexity. Human subjectivity or inwardness is not an anomaly within nature, but according to the Darwinian thesis of evolutionary continuity, must be prefigured in less complex biotic forms. Inwardness is therefore an appropriate point of entry into the phenomenon of living nature; it is that part of nature that we as humans know most intimately. The scale of life's complexity begins with metabolic processes that all forms of life share and ascends through motility and desire, to sensation and perception, and finally to the human mind. In short, the wholesale philosophy of life upon which Jonas's ethics depends will comprise organism and mind both, an integral monism. The human cannot be disentangled from the other forms of life from which it is generated.

Jonas's characterization of his methodology as an "existential" interpretation of bios contains two basic meanings: critical-philosophical and scientific-empirical. By identifying his methodology in the way that he has, it is important to recognize that Jonas is raising a significant challenge to Heidegger through an extension of his categories. It is in this sense that his "existential interpretation of biology" is itself part of a larger substantive *critical philosophical* project. For Heidegger, as already discussed, human "ek-sistence" is fundamentally different from the "existence" of other forms of life. Only the human "ek-sists" or stands out "ecstatically" in what he calls the clearing of Being.[22]

As explained previously, Jonas understands the work of Heidegger not so much as a departure from the history of Western philosophy as the culmination of this history's deeply rooted metaphysical dualism. On Jonas's view, the severance of the human from the rest of nature and of being from value, reflected in Heidegger's categorical distinction between biotic "existence" and human "ek-sistence," is the ground of the nihilism that found its deepest and most horrific expression in the Jewish Holocaust. By separating the human so radically from nature, human life is bereft of naturally and objectively grounded norms to guide it. All value is but an invention of the human will. No value is given in the nature of things. Thus, by extending an existential interpretation to nonhuman forms of life, Jonas aims to reconnect the human and the natural, and to ground moral value objectively in the nature of things. The significance of his methodological critique of Heidegger consists of exploring the way that human capacities for "ek-sistence" do not sever the human from nature but instead arise through natural processes, and therefore must also be present in nonhuman modes of being.

But it is not only that Heidegger's categorical distinction between humans and the rest of life is morally problematic. It is also, as suggested previously, empirically and biologically suspect. To the extent that Jonas can demonstrate this, his "existential interpretation" of bios becomes a *scientific-empirical* project in addition to a *critical-philosophical* one. According to Jonas, while it may be true that only the human is consciously aware of his existence, it is not true that the human alone cares about existence. Heidegger's account of the distinctiveness of Dasein turns on the thesis that the human alone is concerned with

---

22. In his *Letter on Humanism*, Heidegger writes that the "human body is something essentially other than an animal organism." The human essence for Heidegger has nothing to do with human animality. Even if the physical human body is related to other animal life, human being is qualitatively distinct from the animal. The "ek-sistence" of the human is a condition totally incommensurable with all other living conditions. Heidegger writes on this score, "Of all the beings that are, presumably the most difficult to think about are living creatures, because on the one hand they are in a certain way most closely akin to us, and on the other are at the same time *separated from our ek-sistent essence by an abyss*" (for this and the previous, see Heidegger, *Basic Writings*, 228, 230). So for Heidegger, "ek-sistence" and "existence" are fundamentally incomparable and to treat them as similar concepts is a profound categorical confusion.

the question of Being because the human alone is aware of being-toward-death. Dasein's care for being derives from Dasein's cognizance of mortality.[23] On Heidegger's view, this connection between care and mortality marks the "ek-sistence" of Dasein.

But for Jonas the connection between care and mortality characterizes all living entities insofar as, on an account of biotic continuity and the phenomenon of metabolism, all life is to some degree purposive. Even the most primitive organisms care for their own being insofar as they strive against nonbeing. The empirical fact of organic striving, evidenced in metabolism, suggests to Jonas that all organisms exist, to some degree, in purposive relation to their ownmost possibility of death.

At this point, two key meanings of Jonas's characterization of his project as an "existential interpretation of biological facts" have been examined. The *critical-philosophical* meaning refers to a morally and politically driven departure from and challenge to the dualism coursing through the history of philosophy, especially as that dualism is manifest in the thinking of his mentor Heidegger. The second meaning of his "existential interpretation," I have suggested, is *scientific-empirical*. This pivots on Jonas's thesis about the essentially purposive nature of bios. Given evolutionary biology's theory of continuity, whatever capacities exist in human beings must to some degree be prefigured by less complex forms of life. *If* humans are linked through evolution to other forms of life, *then* human capacities must be traceable back to other biotic entities. With these meanings in hand, a description of the phenomenological mode of Jonas's existential biological methodology is now appropriate.

As just discussed, Jonas extends the category "existence" to all forms of life through appeal to life's evolutionary continuity. Through continuity he identifies a basis for reading purposiveness as an attribute of the organic as such, not merely the human. Purposiveness is attributable to any entity whose life consists of a negotiation of certain po-

---

23. As will be analyzed in chapter three, awareness of death, and specifically the memorialization of death through ritual, is one of three central anthropological themes for Jonas. The crucial difference between Jonas's embrace of this theme and Heidegger's is that Jonas interprets this awareness and its ritualization in continuity with the rest of natural life rather than as marking a radical disjuncture. Jonas's view of the anthropological significance of "the grave" has its roots in his interpretation of the phenomenon of metabolism, which will be discussed below.

larities, such as those between "freedom and necessity, autonomy and dependence, ego and world, connectedness and isolation, creativity and mortality...."[24] These polarities, to be sure, are especially recognizable in the human realm. But if the evolutionary theory of organic continuity is correct, they are also present "*in nuce* in life's most primitive forms, each of which maintains a perilous balance between being and nonbeing and from the very beginning harbors within itself an inner horizon of 'transcendence.'"[25] If these polarities are present in other forms of life, then it is sensible, according to Jonas, to begin from the best knowledge available, the self-testimony of human experience. This phenomenological approach to the organic, originating in the human experience of inwardness, is not meant by Jonas to supplant the methods of the biological sciences but philosophically to supplement them.

The statement that life's most primitive forms harbor an inner horizon of transcendence is highly complex. It should first be mentioned that Jonas does not mean here that all forms of life exhibit this horizon of transcendence to the same degree. In other words, biotic continuity does not imply sameness. Rather, according to Jonas, forms of life from plant to animal to human exhibit a scale of ascending complexity. This complexity relates to increasing degrees of mediacy between organisms and their environments. Whereas plant life thrives in contiguity with an environment adjacent to it, animal life depends for its existence on a more perilous negotiation of a distanced environment, and human life reaches a degree of mediacy transcending that of other animals. In sum, plants live in dialectical relation with an adjacent environment; animals with an environment that includes a distanced world; and humans with an environment, world, and further, with the ideas of each of these as well as with the idea of themselves.

Central for Jonas on this score is his claim that the phenomenon of metabolism provides an "Ariadne's thread" that discloses the scale of life's complexity by way of an increasing, but never total, mediation or transcendence of physical need. Metabolism is central to his understanding that life as such contains within itself an inner horizon of transcendence. The purposiveness of organisms and their inner horizon of transcendence, for Jonas, are most basically expressed through

---

24. Jonas, *Phenomenon of Life*, 60.
25. Ibid.

metabolic processes. Metabolism is the process whereby an organism sustains itself as an entity distinct from the environment through material exchange with the environment. The organism is not the same as the material it includes, and yet it is to some degree dependent on matter. It is never identical with the matter it consists of since this matter is only temporarily included in it. So the form of the living organism is not reducible to any instance of its material composition. Thus, the independent unity of an organic form, its identity with itself, consists in a resistance to identity with the matter it is temporarily comprised of. Jonas writes, "[The organic form] is never the same materially and yet persists as its same self, *by* not remaining the same matter."[26] The phenomenon of metabolism marks the finitude of all life as well as the divide between the living and nonliving or the organic and inorganic. When metabolism ceases, the organism dissolves back into the environment.

So, through metabolism, the organic primitively affirms itself against nonbeing and displays an inner horizon of transcendence. The philosophical implication of this, for Jonas, is that even simple forms of life display freedom—"indeed, [metabolism] is the first form freedom takes."[27] Since all organisms metabolize, and metabolism is the process constituting the organism's independence from its environment, all of life manifests some measure of freedom. Metabolism is the ground of independence, independence is freedom from the environment, and freedom from the environment is an expression of the organism's horizon of transcendence. Metabolic process, "the basic substratum of all organic existence," manifests freedom and purposiveness primordially insofar as it is through this fundamental mechanism that organisms affirm their being against non-being. Through metabolism, life "chooses" itself against death. It is the capacity and enactment of this choice, even though it is not a conscious choice, that Jonas refers to in saying that from its very beginning life harbors within itself "an inner horizon of 'transcendence.'"

Metabolism for Jonas is key to the origin and essence of life. The origin of life consists in a "primordial act of separation, [whereby some living substance] detached itself from the overall integration of things

26. Ibid., 76.
27. Jonas, *Mortality and Morality*, 60.

within the totality of nature, positioned itself vis-à-vis the world, and thus introduced the opposition between 'being' and 'nonbeing' into the indifferent assuredness of existence."[28] This opposition between being and nonbeing is the most basic polarity characterizing life. Against Heidegger, the essential paradox of life, not just Dasein, consists in its constant dialectical struggle with the possibility of not being. The freedom from matter marking the difference between the organic and inorganic is tethered intractably to a necessary dependence on matter. Metabolic process is the "precarious independence" of organic entities in their relationship to the matter that is basic to their existence. Metabolism is simultaneously independence from the material environment and dependence on it. Jonas describes this as the organisms' "needful freedom." Through "this dual aspect of metabolism—its power and its need—*nonbeing* entered the world as an alternative contained within being itself."[29] The organic consists essentially of a polarity between being and nonbeing. Life's essence and its greatest contradiction is its' perishability. Jonas writes, "Life is mortal, not although but *because* it is life . . . ."[30]

Jonas's contrast with Heidegger on this point is subtle. Heidegger would never have claimed that only human life is mortal, that only human life consists of the polarity between being and nonbeing. His claim instead is that only human life *cares* about this polarity, is aware of being-toward death, and therefore is capable of being concerned with the question of Being. In contrast, Jonas claims that the *organic as such cares* about this polarity. But Jonas's view does not naïvely claim that all organisms exhibit this care in the same way. Primitive forms of life may in some sense care, but not in the conscious way that humans do. And yet all organisms manifest a concern with being against nonbeing even if it is only physiologically expressed. In other words, Jonas's existential interpretation of metabolism indicates that even simple forms of life, insofar as metabolism harbors within itself a primordial expression of choice, manifest an inner horizon of transcendence. In contrast to Heidegger's terms, for Jonas all life "ek-sists" by standing out to some degree from the material environment, which, simultaneously, all liv-

28. Ibid., 61.
29. Ibid.
30. Ibid.

ing beings depend upon. Life as such cares for itself and purposively strives against nonbeing, though this purposiveness may not always be held in view. For Jonas, then, the primary concept in his existentialist interpretation of biological facts, metabolism constitutes the difference between the organic and inorganic and provides the "Ariadne's thread" for his phenomenon of life.

Conclusion

As claimed in my introductory chapter, Jonas's project can helpfully be identified as a revolutionary moral ontology in which the hermeneutical, fundamental, and normative dimensions reciprocally influence one another. That claim is in part a methodological one concerning the interpretation of ethical theories. But it also relates to my aim to suggest the constructive possibilities of an ecotheological ethics of responsible participation. Towards that end, it will be helpful at the conclusion of each chapter to take stock of where the argument stands with respect to that intention.

The beginning of this chapter shows that the environmental crisis exerts immense moral pressure in Jonas's thought. The opening quote, in which he relativizes the Jewish Holocaust against the global dimensions of ecological catastrophe, illustrates his perception of the environmental crisis as paramount contemporary moral problem. All other moral issues pale in the face of it. The rhetorical coupling Jonas deploys to make this point casts into high relief his sense of the enormity of our environmental challenges. And yet it also signals that the main impetus to his project is not the state of the environment itself, but what he takes to be the deeper source of the environmental dilemma: the moral ambiguity of human power. This is a profound insight, one that needs to be emphasized in any ethical vision adequate to our endangered time.

Jonas's intellectual course from his early concern with Gnosticism through his philosophy of biology and into his ethical and theological writings is directed essentially by a concern with the morally ambiguous character of human power. He is morally motivated by revulsion—revulsion at the slaughter of millions, including members of his own family, following from the egregious misdirection of power in Nazi Germany; and revulsion at the state of "mute things" in the grip of raw human power. In both cases it is not only human power that leads the destruction, but it is humanity that is ultimately threatened. Even more, as will be discussed later, it is not merely the continued physical existence of individual humans that is at stake but the perpetuity of moral responsibility, which Jonas considers to be of the essence of what being human means, and thus the future possibility of humanity as such. The moral impetus backing Jonas's project, then, is the ambiguity of human power and the pressing need, for the sake of humanity and the future of life, to articulate an imperative of responsible power.

An ecotheological ethics of responsible participation endorses Jonas's view, beginning to emerge here, of the need to develop a normative account of the human in response to the environmental crisis, and the position that such an account must place significant emphasis on the native biological roots of responsibility. As I put this earlier, Jonas argues that without objective grounding in the nature of things, in human life and the broader natural world, meanings, values, and norms are no longer envisioned or registered and there is little to constrain the human will-to-power. Power without direction or limitation holds sway; power becomes value's only recourse. According to the historical genealogy I traced in this chapter, for Jonas this defines the ethos of our time and is the essential moral problem to which his project responds. This is a crucial point of departure for response to the environmental crisis, drawing attention to the morally rudderless character of the contemporary human relation to nature.

The normative priority of the human in Jonas's project, catalyzed by his revulsion at the destructive powers of humanity, characterizes his ethical system as anthropocentric. And yet this anthropocentrism is qualified to the degree that his methodology is biologically informed and environmentally attuned. An ecotheological ethics of responsible participation embraces this chastened anthropocentrism. The logic underlying Jonas's qualified anthropocentrism will be detailed in coming chapters, but briefly it expresses the idea that the best protection for the natural world is to direct ethical theory and practice toward preserving a normative conception of humanity, or more specifically for Jonas, toward the ongoing possibility of responsibility. The preservation of a moral idea of humanity entails ensuring the conditions necessary and sufficient to the exercise of responsibility, insofar as on Jonas's interpretation responsibility is of the essence of being human. Sustaining these conditions will not only protect humanity but also the larger natural world. But understanding what these conditions consist in requires, according to Jonas, a thorough reinterpretation of nature and the human place within it.

According to Jonas, the essence of the historical misinterpretation of the human, and the moral dilemmas it produces, consists of a distorted account of the human relation to the natural world. Against this misinterpretation Jonas enlists what he takes to be one of the profoundest implications of Darwinian theory, that whatever capacities are typical of the human species must be prefigured in less complex forms of life. Human continuity with other forms of life is thus established as a basic methodological theme in Jonas's philosophy of nature. And yet insofar as his methodology is shaped by his phenomenological commitment to human experience, this continuity takes on an anthropomorphic caste.

Jonas treats human freedom and purposiveness as an interpretive rubric for the whole of the phenomenon of life and uses this rubric to pry open an account of the organic in general. He claims to find traces of freedom and purpose present in primitive form in the mechanics of metabolism, one of life's most essential operations. In doing this, he identifies as life's uniting theme what has historically been interpreted as rigidly distinguishing the human from other forms of life. Thus, the biotic continuity Jonas describes is held together and qualified by an accent on the free and purposive human will; but so also, the distinctiveness of humans is qualified by the evolutionary history in which we are embedded.

For this reason, Jonas's existentialist-phenomenological biological theory is simultaneously a source of insight and potentially problematic for an ecotheological ethics. Jonas intends for it to serve as a philosophically speculative supplement to a more reductive physicalist reading of life.[31] Strachan Donnely characterizes this intention of Jonas's as philosophically "regressivist": the thesis of biotic continuity justifies a

---

31. On this point, Gereon Wolters writes that, "Although [Jonas's philosophical and existential biology] is complementary to science, Jonas insists that an adequate scientific understanding of the living is as much in need of its philosophical complement as an adequate philosophical reflection is in need of being scientifically informed." Wolters, "Hans Jonas's Philosophical Biology," 90.

"'regressive' method of starting with human experience and underpins his speculative phenomenological interpretation of organic life."[32]

This supplementation of Jonas's needs critical examination. What does it accomplish? It cogently theorizes the human within the fold of the organic, thus challenging the historical dualism between humans and other forms of life. And it suggestively develops the notion that purposiveness belongs to the organic as such, rather than to humans alone. It correctly brings to view an account of the human and other animal forms of life as psychophysical unities. All of these insights are crucial. But in what ways might there still be some distortion in Jonas's representation of the phenomenon of life? I hold that as creative and insightful as his methodological strategy is, it does not provide a view of all that is descriptively and morally significant about bios or human moral experience.

First, however convincingly Jonas makes the case that evolutionary continuity implies that the category "existence" can be extended backward in organic history, this interpretive rubric conceals from Jonas some of the important insights in the biological sciences. By centering on the metabolic agon of the individual organism, Jonas's approach downplays the spatial and temporal interdependencies of life, the "tangled bank" from which this book draws its title. The particular life histories of all individual organisms in whatever specific place in which they may be situated, whether riparian niche or otherwise, are shaped by a longer history of genetic information that links them in a web of interdependence to other organisms and their broader environments.

The core insight of Darwinian theory is that evolutionary change results from the forces of natural selection operating on the variation of traits within a species population. The possession of certain traits conduces to an organism's survival in certain environments. Though members of a species by definition have many characteristics in common, no two individuals are exactly the same. There is generally a great deal of variation within a species. Organisms that possess those traits among the variety of traits within a species that conduce to survival are more likely to survive and thus successfully to reproduce. Through successful reproduction the traits of the parent organism are passed on to its offspring. In this way, environmental conditions "select" for

---

32. Donnelly, "Philosophy, Evolutionary Biology, and Ethics," 155.

traits that conduce to an organism's survival. But the source of variation largely remained mysterious to Darwin. The neo-Darwinian synthesis contributes to the unraveling of this mysterious source of variation in its understanding of the genetic-informational dimension of evolution. This dimension of evolution complexifies the organic drama upon which Jonas focuses.

For Jonas, the drama of life is played out as a largely mono-a-mono engagement between the individual organism and its environment. This is one of the costs of the great gain provided by Jonas's existential focus on organic inwardness. The existential methodology shapes a competitive-antagonistic dramatization of biological facts, echoing the moral structure of the gnostic universe. Jonas focuses on the dramatic inward tensions of the organic individual largely apart from the broader evolutionary tangle. For Jonas, metabolism is an expression of the will to life against nonbeing. The drama of metabolism is the principle theme of life, a primordial death-awareness that, even if not conscious, is the basis for organic continuity. Within this drama, the environmental context of the organism is construed principally as a threat to it, as something that must be existentially negotiated for the purpose of organic survival.

Jonas acknowledges that through the mechanics of metabolism organisms of necessity draw from their material environments, and that they are thus to some degree environmentally dependent. Yet the significance of this acknowledgement is overshadowed by his construction of metabolism as principally an existential struggle of the biotic individual's will to life against nonbeing. As a result, while Jonas aims to establish human continuity with the natural world against the legacy of dualism, his existentialist interpretation of biological facts can be critiqued for erecting a dualism of a different kind. He substitutes a dualism between the organic individual and the larger environment for the historical dualism between humans and nature.

And so while Jonas's existentialism is biologically revised, extended to the whole of the organic rather than to humans alone, it harbors an individualistic or idiocentric emphasis within it. He focuses on a part of nature, the purposive striving of the organic individual, expressed primitively through metabolism, at the expense of a relational account of the organism and a greater sense for the larger environmental whole which limits and sustains every individual's striving. Again, the strengths of Jonas's methodology include the significant uniting of the human and other biotic forms and the speculative phenomenological supplement it provides to a reductive physicalist interpretation. Jonas's existential methodology, in short, shows the deep animality of human beings. And yet by fastening on the inward existential tension within organic individuals, Jonas downplays the relational, symbiotic, and cooperative aspects of life's evolutionary history. Jonas's idiocentric existentialism works like a microscope, helpfully bringing to view a finely grained picture of the organic individual within evolution. However, in the aim of gathering a fuller picture of the phenomenon of life there is need for this microscopic existentialist supplement itself to be supplemented by a wide-angle contextual view.[33]

Philosopher Leon Kass expresses another point of critique. While acknowledging his indebtedness to Jonas's remarkable contributions to the philosophy of biology, Kass believes that Jonas's existentialism leads him to neglect the significance of organismic sexuality and reproduction.[34] A phenomenon of life leveraged on metabolism yields a picture of the world comprised of individual organic competitors in a threatening environment. Emerging through an emphasis on sexual desire and the organic drive to reproduce, on the other hand, is a picture not of centripetal, autonomous struggle but of centrifugal, relational complementarity. These are fundamentally opposed images of life. In any effort to construct an ethical theory rooted in a descriptive account of life, these oppositions must be seriously considered.[35] And yet according to Kass, Jonas's neglect of the relational-sexual dimension of the phenomenon of life reflects the degree to which he may be following rather than critiquing Heidegger by "homogeniz[ing] the outer world and exaggerat[ing] its loneliness."[36] While Jonas does effectively challenge Heidegger's dualism between human and other forms of life, in erecting a dualism between the organic and the environment he transposes the existentialist view of the indifference of nonhuman nature onto the

---

33. Again, Strachan Donnelly writes, Jonas's methodology "may underplay the fact that life's forms, capacities, and individuals arise in populational and ecosystem communities, dynamically, interactively, and relative to one another in a constant process of biotic variation, diversification, and selection" (Ibid., 137).

34. For this critique, see Kass, "Appreciating *The Phenomenon of Life*," 3–12.

35. The account of responsible participation I am working toward intends to reflect the significance of these oppositions at the level of a moral anthropology.

36. Kass, "Appreciating the *Phenomenon of Life*," 11.

inorganic environment. Thus one of the strengths of the hermeneutical dimension of Jonas's project is to shrink the domain of the indifferent other, but one weakness is his under-theorization of life's pronounced relationality, interdependence, and complimentarity.

This concluding critical summary of the hermeneutical dimension of Jonas's thought is not meant to detract from the significance of his effort to revise the historic ideological distortions of the human relation to nature, but to suggest that as creative and persuasive as this effort is, it still provokes some important questions. The aim of my larger project is to draw from both Jonas and Gustafson and to advance the significance of their work by moving beyond some of their flaws. Jonas's phenomenology of life importantly focuses on the metabolic drama of life. This is a crucial point of linkage between humans and the rest of life, and belongs within any comprehensive moral anthropology and description of natural life. But Jonas's commitments to organic autonomy and competition yield an incomplete picture of the phenomenon of life, and as will be shown in chapter five and my conclusion, are at the source of his most significant flaw. It underwrites a moral anthropology of responsibility that provides only a partial image of the fuller dialectic of human experience in which humanity indeed uniquely stands out from, but also participates within, natural processes. This interpretation of the hermeneutical dimensions of Jonas's work, the moral motivation backing it and the methodological commitments informing it, suggest some crucial points of contrast and complementarity with Gustafson. The following chapter moves to an interpretation of these.

2

Divine Power and Critical Religious Naturalism

Introduction

IN THE PREFACE TO HIS MONUMENTAL TWO-VOLUME *ETHICS FROM A Theocentric Perspective*, Gustafson affirms that his book is the "product of at least thirty years of 'homework' and fifty-five years of living."[1] It is not only a product of scholarship but also of reflection on the events and circumstances of his life. The effort to understand thinkers such as Gustafson and Jonas requires that some attention be trained on their biographies.

For Jonas, one of the life-experiences that so radically shaped the direction of his work was that of World War II. Removed from his books and seminars, living in the trenches of resistance against his homeland and the barbarism of Hitler, Jonas became acutely aware of the perishability of life. This awareness and concern shaped the trajectory of his thought and scholarly output from that point on. For Gustafson the relevant life-experiences occur much earlier and are more subtle and diffused.

Gustafson's account of the impact of his early formative experiences registers a more complex view of the human relation to nature and history and the divine in comparison to Jonas's more focused attention on organic individuals.[2] But like Jonas, Gustafson understands

---

1. Gustafson, *Ethics from a Theocentric Perspective*, 1:ix.

2. One of my burdens in these first chapters is to draw out the significance of the relatively greater relationality of Gustafson's moral anthropology in contrast to Jonas's accent on individual autonomy. In doing this, however, I do not mean to suggest that Jonas's anthropology does not include a relational dimension. But the relation emphasized in Jonas's work is that between a relatively autonomous individual and, as suggested in the conclusion to the preceding chapter, a somewhat homogenized account

that any human experience of the natural environment is conditioned by "decisive personal experiences that have affected our attitudes and outlooks."[3] Crucially for Gustafson's methodology, as I will draw out soon, "experience" precedes explanatory rubrics, intellectual and moral spheres of attention, religious sensibilities, and the cognitive, affective, and evaluative influences of traditions of thought and practice.

In *A Sense of the Divine*, focused specifically on an environmental ethical application of theocentrism, Gustafson acknowledges the formative impact of his childhood in the northern woods of Michigan's Upper Peninsula. Through long walks in the woods he learned to identify the distinctive flora and fauna of his home place, he relished foraging for berries and nuts. He canoed on nearby lakes and rivers, helped to chop and pile wood for his family's kitchen stove, and delivered the local paper in the extreme winter weather conditions of the far north. In addition to these positive experiences in his home environment, he was also deeply affected by a local paper mill's pollution of the Menomonee River, which ruined the fishing down river, and by the ugly heaps of slag from a nearby iron ore mine. When his family later moved to Kansas, he experienced and observed the devastation of a tornado and the infamous drought in the American plains of the 1930s.

In recounting these experiences, Gustafson acknowledges that he was early struck with awe by both the grandeur of nature's beauty and nature's great power to harm and destroy, and the way in which this awe is correlated to his enduring questions about the nature of God. Thus, in contrast to Jonas's concern with the moral ambiguity of *human power*, Gustafson's moral concern can helpfully be characterized more broadly as a theological concern with human moral evaluations and conceptions of *divine power* as these relate to understandings of the human good.

Gustafson's awe before the power of the divine funds his scholarly agenda and contrasts with Jonas's focus on human power. Where Jonas's motivation is principally practical and anthropocentric, Gustafson's

---

of the natural environment. As I put this earlier, Jonas's vision of the relation between organisms and environments can be understood basically as an agonistic one. In contrast to this, the relationality of Gustafson's anthropology comes to view as a more highly textured reciprocity between human selves and communities, the patterns and processes of nature and history, and God.

3. Gustafson, *A Sense of the Divine*, 1.

is quite different. He is motivated by a concern with the ambiguous patterns and processes of the world, natural and historical, which are ultimately divinely governed, and with the conceptual and practical challenges of morally evaluating this ambiguity from a theological ethical perspective. In this chapter I will interpret Gustafson's accounts of our morally troubling circumstances and the root cause of these, move to a treatment of his methodology, and conclude with a critical analysis of his hermeneutics.

The Problem of Anthropocentrism

Gustafson's interpretation of our cultural situation centers on the problem of anthropocentrism. He is motivated by a concern with the way in which anthropocentrism, which he deems to be suspect on scientific and theological grounds, exacerbates the human threat to the natural world and instantiates what he takes to be the fundamental flaw of human beings. With Jonas, Gustafson too is concerned with the powerful consequences of contemporary human action and holds that the scale of modern human efficacy today is distinct from any other time and unleashes new moral challenges.

The expansion of knowledge and technological innovation frees modern humans in many important ways from historic insecurities in relation to the natural world. For example, medical advances, pharmaceutical sciences, and the sciences of nutrition and sanitation have lengthened average human life spans and increased life-quality for many. And along with Jonas, Gustafson also affirms the biological underpinnings of unique human capacities. The history of biological evolution has resulted in the development of human capacities that extend "the range of [our] domination over forces and powers" that are beyond the control of other forms of life. A result and feature of these distinctive human capacities is the creation of culture, a second nature of "artifacts and meanings which are shaped to render the life of human community more immune to the uncertainties of natural conditions . . . ."[4]

And yet, no reflective person need be told that the extension of human mastery not only has not fully eliminated certain fundamental and unavoidable anxieties, but that it has introduced new anxieties. While rendering humans more secure in relation to certain contingencies,

---

4. Gustafson, *Ethics from a Theocentric Perspective*, 1:4.

new threats have also been created. Gustafson writes, "this increase in mastery has not eliminated insecurity and anxiety: these feelings are evoked by different objects, by other contingencies, including new ones that are the unintended and unanticipated consequences of the extension of human mastery itself."[5] Gustafson here shows that he is sensitive to Giddens' distinction between "external" and "manufactured" risk, discussed in my introduction, and that his interpretation of contemporary circumstances corresponds in a basic way with Jonas's.

But while consonant at a basic level, Gustafson's critique of our circumstances differs from Jonas's in two important respects, rhetorical and substantive. Rhetorically, Gustafson's writing is much less prophetically edged than Jonas's. As Robert Bellah notes, Gustafson's writing is a "measured reconnaissance," undertaken at a "leisured pace."[6] This rhetorical calm of Gustafson's writing contrasts sharply with the revulsion driving Jonas's. Where Jonas writes like a Jeremiah, Gustafson manifests the detached wisdom of a Solomon. Gustafson's muted, leisurely reconnaissance is perhaps ironic considering the radical challenge his theocentrism poses to the historically dominant anthropocentrism of much of Western theology and philosophy. But this may not be the case. The rhetorical differences between Gustafson and Jonas are not merely superficial, but reflect significantly different perceptions of the gravity of the environmental crisis of power and of the right moral responses to this problematic.

Differently from Jonas, Gustafson's critique of the contemporary ethos is explicitly theological. In supplement to Jonas's idea that the unanticipated and unintended results of contemporary power are inherent to the processes and tempo of modern technological innovation, Gustafson's theological judgment is that this is shaped ultimately by the inwardly curved nature of human individual and communal self-interests. He considers this ultimate cause, what he calls the fault of contraction, to be the fundamental human flaw, the sinful condition of human nature.

---

5. Ibid., 5.

6. Bellah, "Gustafson as Critic of Culture," 143. The volume in which Bellah's article appears contains a collection of papers from various distinguished interpreters of Gustafson given at a symposium on his *Ethics from a Theocentric Perspective* held at Washington and Lee University in 1985.

By "contraction" Gustafson designates the anthropocentric narrowness of a human sense of the world that stems from an exaggerated sense of human significance within it. All human achievements reflect this contraction insofar as all achievements are motivated by human valuations that are always perspectival. "Increase in knowledge is a purposive activity," Gustafson affirms, because "it stems from the valuations of individuals and communities; it is directed toward ends that human beings value."[7] Given that purposiveness directs innovation, one can reason deductively from innovation to what humans in fact value, for example quality and length of life or material prosperity. But this, of course, is not at all the same as being able to determine *why* humans value what they do or *what humans ought* to value.

Why humans value what they do tends not to receive sufficient critical scrutiny. This, according to Gustafson, is partly the effect of the tendency of human valuations to be "curved in upon our immediate self-interests."[8] For Gustafson, humans are in denial about and neglect critically to face the inward curvature of our valuations. To the extent that this is the case, he claims, we refuse to acknowledge our human limitations and this refusal is to human detriment and the planet's, along with being idolatrous. Gustafson's rhetorical style, always carefully nuanced, reflects the impact of this judgment on his own scholarship.

Gustafson writes, "Knowledge and foreknowledge are expanded; capacities for control of future events are extended; finitude, however, is not overcome."[9] The denial of finitude is at the core of Gustafson's theocentric indictment of anthropocentrism. According to Gustafson, this denial is inherent to anthropocentrism and foments the transgression of human limitation and the limits of the natural world. While the character of "limits" transgressed, the modes of critique of such transgressions, and the proposed remedies for it are culturally and historically variable, Gustafson argues that the sense that there are appropriate human limits that need to be observed is philosophically and historically perennial.

Critiques of the failure to acknowledge human limitation and finitude extend back through all recorded history. For example, ancient

---

7. Gustafson, *Ethics from a Theocentric Perspective*, 1:7
8. Ibid., 8.
9. Ibid.

Greek tragedians and philosophers understood this as hubris and the biblical traditions as pride and sin. For both, the tendency to exaggerate human powers results in a distortion of humanity. While the tendency to pride and hubris is ancient, perhaps an intractable element of the human condition, the necessity of resisting it is in Gustafson's judgment more urgent in this time than ever before. Along with Jonas, Gustafson affirms that this is due to the fact that human efficacy has increased to the point that the moral consequences of pride and hubris now have a planetary dimension.

In contrast to Jonas, however, properly theorizing human power for Gustafson is a theological task that must be developed, in part, through analysis of religious experience. This demand is of course not based on the view that religious practitioners always have a proper sense of their place in the world or of their relation to God. Within religious life, the failure to acknowledge limitation is often reflected in the instrumentalization of religious beliefs and practices. According to Gustafson, religion is appreciated today almost solely for its utility value in the quest to secure temporal human ends. Religions on this account are reduced to therapies. The general religious admonition to "live for others rather than for self is, in modern interpretation, a cause of the problem [of individual self-fulfillment] rather than an answer to it . . . ."[10] On this popular therapeutic account, binding obligations and duties to others, not to mention to the divine other, corrode rather than conduce to happiness. But for Gustafson there is much more to the phenomenon of religious conviction than its conciliatory character.[11]

At the core of Gustafson's project, then, is the burden to offer a non-utilitarian account of religion. Whatever values or consolations the religious life offers, he argues, ultimately stem from consent to the divine rather than to the finite, limited, inwardly curved interests of human selves and communities. Such consent reorders human interests and demands an enlargement of moral concern. Pride and hubris result from the failure to give this consent and shrink the horizon of moral sensitivity. Immediate self-interests take precedence over the divine will and the long-term goods of self, others, and the larger world. In Gustafson's judgment, "contemporary instrumental religion is wrong

10. Ibid., 20.
11. Ibid., 31.

theologically as well as practically because it does not set human life within the appropriate limits, not only of finitude, but of ordered relationships in institutions and between persons."[12] To redress this, Gustafson understands the task of his constructive work to be that of resituating the human as a participant within the appropriate limits of ordered relationships.

In sum, Gustafson's interpretation of our cultural circumstances consists of a critique of the entrenched anthropocentric view that the human species is the measure of all things. He does not deny that humans are endowed with unique capacities for thought and action. He does not deny that humans alone exist as "measurers" of the world by way of these unique capacities. He does, however, challenge the tendency to collapse the difference between "measuring" and "measure." That humans can interpret, evaluate, and influence the world in ways that other forms of life cannot does not necessitate the position that the human species is the center of value in the world. Collapsing this difference is the essence of anthropocentrism.

In contrast to Jonas, the correction to anthropocentrism cannot depend on mere revision or qualification. As I will show further along through my interpretation of the fundamental and normative dimensions of his work, Gustafson challenges anthropocentrism by articulating a theocentric construal of the world and the implications of this construal for the moral life. Among other important things, this challenge consists of a critique of moral anthropologies emphasizing autonomy and a call for a heightened awareness of the dependencies and interdependencies that shape our personal and social lives, our relations to the natural world, and above all our relationship to God. But in order to clear the path for examination of these claims, it is important first to turn to an analysis of the methodology that supports them.

Critical Religious Naturalism

Gustafson's project is shaped by a very broad theological ethical question: "what is God enabling and requiring us to be and to do?" To this question, he brings to bear a complex, variegated interpretive methodology, best characterized as a critical religious naturalism. He combines an emphasis on experience with a critical respect for its mediation by

12. Ibid.

tradition, theological, ethical, and scientific. While the character of his thought is emphatically theological, it is generated by reflection on the common, natural human experiences that underlie religious sensibilities rather than from prior theological or ethical premises—what I construe as a critical religious naturalism. He shares Jonas's phenomenological commitment, but for Gustafson this funds a description of the world whose core insight is that there are forces and powers bearing down upon and sustaining life that are beyond human control, and these can be and often are religiously construed. From this, the task of the moral life becomes that of discerning appropriate practical responses to these powers. In the following sections of this chapter, I will first examine Gustafson's methodology in its broadest character, and then I will transition to a closer analysis of the various focal commitments within the methodology.

A "Composite Rationale"

One of the distinguishing marks of Gustafson's phenomenological approach is what Stephen Toulmin characterizes as the simultaneously "old-fashioned" and "revolutionary" theological discourse through which he relates it. It is old-fashioned, according to Toulmin, to the extent that it is stridently unapologetic. Gustafson writes and thinks in an uncompromisingly theological mode, he is secure in the justification and communicability of his Reformed theological interpretive lens. He is utterly convinced of the moral relevance of belief in God. And yet, perhaps ironically, it is this "old-fashioned" theological discourse itself that contributes to the "revolutionary" character of his theological positions. According to Toulmin, "Because [Gustafson] refuses to compromise the claims of his religion or explain them away as fictions or metaphors, he faces head-on issues that his contemporaries are prepared to fudge: notably, issues that arise out of post-Reformation changes in our scientific views about the world."[13] Gustafson does not shy away from but vigorously engages the broad fields of the various Western philosophical and scientific traditions precisely because his "old-fashioned" theological point of view demands this.

13. Toulmin, "Nature and Nature's God," 37–38.

More explicitly than Jonas, Gustafson's methodology is informed by a concern with the degree to which descriptions, explanations, and evaluations reciprocally influence one another. He is highly cognizant of the fact that no data are unmediated, that the tasks of description and evaluation are not purely isolable and always impinge on one another. Description is always partly influenced by pre-reflexive affective responses even as these affective responses themselves are always at least in part organized and codified by inhabited traditions of thought. It is this understanding that leads to the "critical" character of his religious naturalism that so significantly contributes to the fundamental and normative aspects of his work. With respect to his fundamental ideas, Gustafson's critical religious naturalism leads to a twofold focus on, first, the historical, natural, and cultural contexts of human moral existence, and, second, on the centrality of affectivity in his moral anthropology.

Like Jonas, Gustafson is an interdisciplinary thinker. The justification for Gustafson's interdisciplinarity rests on several mutually enforcing presuppositions. *Ontologically*, he presupposes significant unity between religious and scientific realities. Given this unity, support is granted to an *epistemological* presupposition that various modes of inquiry, such as religious and scientific, can gain access to dimensions of this common reality. Underlying and integrating these ontological and epistemic presuppositions is the *theological* conviction that "the experience of God's reality within the context of the Christian community and tradition is multidimensional."[14] This theological affirmation of the multidimensionality of Christian religious experience means that "any articulation of that experience [of God] . . . must take into account the various aspects of God's relation to man that are present."[15] Any effort to make the experience of God intelligible, in other words, demands that the broadest spectrum of relevant explanatory and interpretive resources be deployed. Thus, for Gustafson, scientific, theological, and ethical work mutually support one another and are necessary to gaining comprehensive understanding of human moral life in the world before God. As one of his interpreters has put this, Gustafson's project is ad-

---

14. Gustafson, *Can Ethics be Christian?* 138.
15. Ibid.

vanced through the deployment of a "composite rationale" that includes attention to experience, the sciences, the Bible, and tradition.[16]

Gustafson's interdisciplinarity is clearly influenced by his mentor H. Richard Niebuhr's understanding of Christian moral philosophy.[17] The pluralistic methodology Gustafson deploys, supported by his view of the porous boundaries between the sciences, theology, and ethics, has been interpreted as both a great strength and potential weakness of his work. Within Christian theological circles, there are both critics and champions of Gustafson. Critics challenge Gustafson's view that common human experience and the sciences should be the significant criteria for the backing and revision of traditional Christian doctrines. The basic charge is that his use of the sciences is excessive and leads to the abandonment of central tenets of the tradition, such as classic Christological, soteriological, and eschatological doctrines.[18] Above all, in reference to his most systematic and comprehensive work, *Ethics from a Theocentric Perspective*, critics ask whether there is a sufficient distinction made between "God" and "Nature" as the ultimate reference of reality.[19] According to this line of questioning, it is not ultimately clear how his position differs, except insofar as it incorporates modern scientific insights, from the classic Stoical position that the moral life consists of piously reverencing and conforming to the logos of Nature.

Yet along with critics there are also many who affirm the methodology that shapes Gustafson's project. Writing on the positive contributions of Gustafson, theological ethicist Harlan Beckley, for example,

16. See Reeder, "Dependence of Ethics," 119–37. This "composite rationale" is roughly equivalent to what has come to be known as the Wesleyan Quadrilateral.

17. As Niebuhr explains in *The Responsible Self*, by describing himself as a "Christian moral philosopher" he intends to distinguish his work from Christian ethics understood narrowly as moral reflection on the particularities of Christian life. In other words, Niebuhr's object, as with Gustafson's, was not the Christian moral life in particular but human moral existence more broadly, considered from a Christian theological perspective. The task of Christian moral philosophy, as Niebuhr articulated it, is to think philosophically from a Christian perspective about human moral existence in general. This justifies and requires the use of multiple interpretive and explanatory resources in addition to specifically Christian ones, insofar as these are crucial to understanding the human moral situation broadly.

18 See for example, McCormick, "Gustafson's God," 37–52. I will deal at more length with these critiques in chs. 4, 6, and my conclusion.

19. See the essays by Gordon Kaufman, Edward Farley, John H. Yoder, and Robert Audi in *James M. Gustafson's Theocentric Ethics*.

argues that, "Christian ethicists will more adequately exercise their vocational responsibility to persons within and outside the church and to society insofar as they adopt something similar to Gustafson's approach."[20] Arguments such as this turn on an expansive idea of the Christian scholar's vocation. The Christian scholar's proper vocational responsibility, on this view, extends beyond the boundaries of those within the Christian tradition to engagement with non-Christian and even non-religious people.

To the extent then that Gustafson's interpretation of what is going on in the world is interdisciplinary, he lives up to the broad mandate of the Christian scholar's responsibility to engage the broader public. The justification for this broad engagement is to clarify through interdisciplinary engagement the general question of what is going on in the world in order better to offer general and specific prescriptions for how morally to navigate human life within it. The project to clarify ethos and to offer normative guidance is best served, so this line of reasoning goes, when various disciplinary perspectives are brought to bear on it. As Gustafson's mentor, H. Richard Niebuhr put it, the general concern with the interpretation of ethos is a philosophical common denominator that allows the Christian moral philosopher to speak beyond the distinctive questions of his or her own community to the questions of human moral existence in general. Thus Gustafson's concern with the common denominator of human moral experience allows him to position himself as an interlocutor in the wider arena of philosophical and ethical discourses.

In the following sections of this chapter, I will provide summaries of the main elements of the composite rationale of Gustafson's critical religious naturalism. Attention will be given, first, to his understanding of the methodological primacy of experience. After this, I will move to his interpretation of basic religious senses and his philosophy of religion. And then I will critically examine his understanding of theological symbols and his account of the proper relation between theology and the sciences.

---

20. Beckley, "A Raft that Floats," 203.

Experience

Gustafson's methodological commitment to the priority of experience in some respects parallels Jonas's phenomenological-existentialist biological methodology. Jonas, recall, interprets the Darwinian thesis of evolutionary continuity to mean that the inwardness and purposiveness characteristic of human life can be deployed as an interpretive key to the phenomenon of life generally. For Gustafson, the relevant category is not inwardness but a complex rendering of experience that grants prominent attention to its contextual conditions. He takes experience to be prior to reflection. And yet, experience is always embodied and relational, an effect of the self's encounter with the external world. Linguistic and conceptual constructions proceed from experience.

Gustafson treats *religion and morality* as first-order dimensions of experience and *theology and ethics* as second-order critical reflection and construals of experience. In other words, theology and ethics are explanatory, interpretive, and theoretical filters of primary religious and moral human experiences. Given this phenomenological caste of his methodology, he acknowledges that some degree of anthropocentrism is unavoidable. But this methodological anthropocentrism, he argues, does not yield an axiological anthropocentrism. Like Jonas, Gustafson grants human experience methodological priority, but rejects the axiological extension of this methodology.

What more specifically, though, does Gustafson mean by "experience"? Differentiating experience into, for example, aesthetic, religious, and moral dimensions can be helpful in specifying this, but in order not to obscure the integrated nature of experience, Gustafson argues that this differentiation should not become overly rigid. The various dimensions of experience commingle and cannot be captured neatly by concepts and analysis. This goes not only for the dimensions of experience, but also its modes. For example, ethical theories often prioritize cognitive and volitional modes of moral experience over the affective. Analyses that focus on either the will or intellect, that is, ethical theories that are primarily rationalistic or voluntaristic, are often used to delineate the moral from the nonmoral. But according to Gustafson such delineations tend to diminish the density of moral experience. The typical prioritization of the intellectual or volitional modes of moral experience stems from the judgment that if the concern of morality is the action of human agents and "agency requires both the formation of purposes or intentions and the direction of the 'will' by those purposes," then "the key [i.e., morally salient] features of human subjects are asserted to be cognitive-intellectual and volitional."[21] The affections on this reading are interpreted as premoral conditions of the capacity for moral action. But as the aesthetic, moral and religious dimensions of experience resist absolute distinction, so also for Gustafson should moral experience not be overly fragmented into cognitive, volitional, and affective modes.

Thus, according to Gustafson, too rigidly differentiating the modes of moral experience falsifies "the unity of the human self, and lead[s] to an underestimation of the interrelations and interpenetrations of the capacities [i.e., modes] conceptually distinguished."[22] The affections, will, and intellect cohabit within agents and structure the variegated quality of moral experience. The good, for example, must be intellectually apprehended in some way in order potentially to be willed. The affections influence both the activity of the will in response to that which is intellectually apprehended as well as the focused direction of the intellect toward certain objects rather than others. So, for Gustafson, the modes of moral experience interpenetrate. Whether moral or religious or aesthetic, a full account of the dimensions of experience is not reducible to any one of these modes apart from the others.[23]

In addition to prioritizing experience and articulating its complex character, Gustafson affirms that experience and its interpreted meanings are inevitably social. However, this does not mean that experience is only socially constituted. Social interaction is a necessary condition of human development and the meanings of experiences are articulated through communal signifying systems. Thus the social character of language through which experiences can be shared and through which their meaning is articulated provides a further communal mediation of experience. The adequacy or inadequacy of the meaningful construal of experience varies according to the degree to which individuals agree

---

21. Gustafson, *Ethics from a Theocentric Perspective*, 1:118.

22. Ibid.

23. Jonas, as the previous chapter signaled and as further chapters will develop, grants analytic priority to the will in his account of moral experience. This commitment is therefore open to critique, on Gustafson's account, insofar as it can tend to reduce the texture of moral experience by fastening on a single dimension of it.

with the convictions of the community that generate those construals. Because of the social character of experience and meaning-construal, explanations and interpretations of experience are historically changing and "any organized community that freezes its requirements for membership according to the symbols and explanations adequate at a particular time is bound to have difficulties."[24] Though an historical fact, this fluctuation of symbolic meanings presents the problem of epistemological relativism.

The Enlightenment quest for a universal rationality and an invariable morality is motivated in large measure by recognition of the problems of the historical and social variability of construals of meaning and value. Along with a now familiar multitude of critics of these Enlightenment aims, Gustafson claims that neither a universal language of reason nor a universal moral calculus is possible. But he adds that this limitation does not require resignation to either epistemological or ethical relativism. Within and between communities and traditions, construals of value and the meanings of experience are testable. Testability depends, in part, on Gustafson's conviction that experience is always relationally responsive to some objective, extra-linguistic, trans-communal dimension of reality. Experiences and their socially interpreted meanings are always elicited or evoked by "the presence of that which is objective to individuals or to communities . . . ."[25] This conviction about the existence of an external, objective presence means that, "we can engage in 'reality checks.'"[26] Judging the adequacy of an interpretation of the meaning of experience depends not only on whether or not an individual subscribes to the convictions of the community bearing that interpretation. Judgment can also turn to a trans-communal assessment of how that interpretation matches up with the object to which it is a response.

The argument that experience is primary, that it is social-relational, that meaning is communally articulated, *and* that judgments of interpretive adequacy can turn to standards beyond community is a difficult argument to sustain. To what trans-social criteria of testability can one appeal to establish the validity and veracity of one's experience,

---

24. Ibid., 124.
25. Ibid., 128.
26. Ibid.

whether religious or otherwise? Gustafson's response to this question is shaped by a philosophy of religion I construe as a critical religious naturalism, to which I now turn.

Religion

As just stated, Gustafson is trying to make a very difficult argument. He claims that experience is primary, that its expression is communally mediated, and also that it is possible to make trans-communal judgments about interpretations of experience. Though difficult, Gustafson aims to show the viability of this claim by grounding it in a critical religious naturalism comprised of an account of religion's underlying, common, natural senses.[27] According to Gustafson, several common human senses underlie every religious tradition's particularized expression of them, though they can also be held together non-religiously in what he considers a stance of natural piety.

> 27. Especially here in Gustafson's account of religious experience it is important to acknowledge the influence of Schleiermacher on his thinking. Of this influence, Gustafson writes, and I quote at length:
>
>> The "empiricism" ... in [Schleiermacher's] work is not unlike [my own account] of how religious affections emerge. The relation of theology to religion, that is, the claim that religion is prior to that theology, and that theology is a reflection on what Schleiermacher calls the religious self-consciousness, is clearly similar. The powerful sense of the divine governance, not only through historical events, but also in and through the nexus of relationships in and with nature is congenial. That a place for human agency is retained without leading to exaggerated claims for freedom is important both to Schleiermacher and to me. While the sense of dependence is the first in my ordering ... it has a dominance in Schleiermacher which I would modify (*Ethics from a Theocentric Perspective*, 1:178).
>
> These noted affinities between Gustafson's and Schleiermacher's understandings of religion will become more apparent in what follows. I would like to note, especially regarding Gustafson's last claim in the quote above, that I will eventually argue that I think one of Gustafson's central weaknesses is precisely the dominance of "dependence" in his own schema of the universal senses and religious experience. Further, though there are deep affinities between Gustafson and Schleiermacher, it is helpful to keep in mind Jeffrey Stout's argument that while Schleiermacher's view of the relation between theology and feeling leads him to call for a "truce" between theology and the sciences, Gustafson's project is clearly driven by the aim to integrate scientific knowledge and theology. And this drive is focused precisely on his understanding of religion as an "affective" phenomenon that is resistant to the distinction between cognition and emotion, or between knowing and feeling that is central to Schleiermacher's understanding of the proper domains of theology and the sciences. See Stout, *Ethics After Babel*, 178–83.

That these senses are common means for Gustafson that all tradition-specific expressions are open to trans-traditional criteria of testability, for any religious construal can be judged with respect to how persuasively it expresses them. This is strategically important to his larger project of demonstrating the broad viability of a theocentric ethics. The natural senses are the platform upon which he develops his philosophy of religion, an understanding of theology, a theory of symbolism, and ultimately his account of how the Christian theological tradition can and ought to include the insights of the sciences.

The potential for religious experience for Gustafson is an effect of the relational quality of human experience in general. The experience of being a human self, for Gustafson, is fundamentally relational. One experiences oneself inwardly always as a self in relation to a world of other selves, natural and historical processes, and for theists, God. Insofar as being a self-in-relation is basic, and gives rise to the universal, common senses that inform religious experience, Gustafson asserts that, "Religion is not unnatural."[28] Religion is not an "idiosyncratic projection of human imaginations untested by experience," but arises from the common experience of being a self-in-relation and the senses correlate to this experience.[29]

These of course are strong claims. They reflect what I am referring to as Gustafson's critical religious naturalism. This philosophical position on the natural bases of religion will prove pivotal for his understanding of the role of the sciences in theological reflection. It is important to note at this point that a naturalistic view of religion enables him to argue that while religions of course vary in many significant ways, they all share in common certain senses that arise from the recognition that humans exist in relation to some things that are beyond human control.[30] His methodological commitment to this religious common denominator provides a critical platform for opening particular religious expressions of this experience, including his own Reformed Christian tradition, to trans-communal criteria of testability.

---

28. Gustafson, *Ethics from a Theocentric Perspective*, 1:134.
29. Ibid.
30. Ibid.

Gustafson enumerates six universal senses: dependence, gratitude, obligation, remorse, possibility, and direction.[31] Dependence is a primal sense that all selves share. No person ever chooses when and where to be brought into the world. All humans are initially dependent on parents or caretakers. All are for the duration of their lives dependent on the sustaining functions of natural and cultural conditions, on powers beyond their control. No self is ever fully autonomous or self-sufficient, but always exists within a network of relations of dependence, both natural and cultural. The sense of gratitude too is common. The relations that we depend upon are usually, though not always, beneficial to us. While many diseases and accidents are beyond human control and tragically impact many people, Gustafson affirms that the "experience of most persons most of the time is that it is better to be than not to be."[32] The experience of the goodness of being and the beneficence of the relations of dependence that sustain human life gives rise to the sense of gratitude.

And again reciprocally, the sense of obligation arises from the sense of gratitude to those others, objects and conditions upon which human life depends. An individual's recognition of dependence, and gratitude for the sustaining network of relations upon which life depends, yields the sense of being obligated to contribute to the work of sustaining those nourishing relations of dependence for others. But humanity often fails to perform the particular actions necessary to sustain beneficial relations of dependence, actions prescribed in various specific ways by the sense of obligation. This failure gives rise to the sense of remorse. No self in all cases is capable of successfully meeting all the requirements evoked by the sense of obligation. Thus the sense of remorse too is natural to human experience. But while never successful in all cases, humans generally sense the possibility of their power and freedom to shape the conditions of dependence for good. And so humans also experience a basic sense of direction. As Aristotle acknowledged, all human actions tend to aim for some end. This end can of course be articulated in multiple ways, both in vague and more specific terms, but Gustafson affirms that, "To be human is to have pur-

---

31. Gustafson's theorizing about these senses is taken up earlier in his career in ch. 4 of *Can Ethics be Christian?* as well as in ch. 3 of *Ethics from a Theocentric Perspective*, vol. 1.

32. Gustafson, *Ethics from a Theocentric Perspective* 1:131.

poses; to have purposes is to intend ends; to intend ends is to have a sense of direction."[33]

These universal human senses of dependence, gratitude, obligation, remorse, possibility, and direction constitute the common contours of human experience, and can be construed in religious and non-religious terms. As universal they are held in common by all humans, making theological and religious dialogue with natural and social scientific accounts of human experience possible and potentially mutually enriching. And insofar as they underlie various religious sensibilities, the various religions share some basic points of contact, allowing for cross-religious dialogue. Gustafson's position on the universality of these senses is at the core of his critical religious naturalism.

That religion is grounded in the six universal senses of dependence, gratitude, obligation, remorse, possibility, and direction, leads Gustafson to suggest that religion is best characterized as basically *affective*. It is not, however, affectivity per se that determines the religious sensibility, but how the object of affectivity is articulated. Specifically religious affectivity is constituted by senses, attitudes, dispositions, and emotions toward objects or powers in the world taken to be ultimate. The ordering of affectivity toward a source of ultimacy distinguishes a religious from a non-religious sensibility.

Gustafson advances his definition of religion as affective by arguing that it can circumvent two historically problematic tensions, the tension between reason and emotion on the one hand and between reason and faith on the other. For Gustafson, affectivity is comprised of senses, attitudes, dispositions, and emotions. In keeping with his emphasis on the multidimensionality of experience and his resistance to analytical reduction, the aspects of affectivity interpenetrate. Gustafson expresses this interpenetration, for example, when he claims that the basic senses are always "senses of" something. In other words, they are object-referential. Thus, his understanding of "sense" is close to what is meant by "consciousness of" something. But to the extent that the term "consciousness" tends to rely on a cognitivist meaning, "sense" moves beyond "consciousness." The various senses as he describes them and the fact that they are always senses of some object or objects means that

---

33. Ibid., 134.

sense includes a level of cognitive awareness. But it also harbors the influence of the other non-cognitive dimensions of affectivity.

Affectivity, for example, includes "attitudes" in addition to senses. By "attitude" Gustafson intends an existential stance or posture toward some object or objects. Attitude carries within it, too, the meaning of having a relatively constant "disposition" toward some object. The reference of disposition is toward an orientation or proclivity to be and to act in certain predictable ways. Dispositions pre-shape actions and account for the relatively stable or settled character and direction of acts by individuals and communities. With the concept of "affections" Gustafson refers to those emotions that are evoked in relations to certain objects. But he is clear to state that the emotional nature of the affections is neither non-rational nor purely subjective.[34]

Through the density of his interpenetration of affectivity, Gustafson claims to move beyond the conflict between reason and emotion in many understandings of religion. As any reduction of experience to one of its dimensions or modes falsifies the unity of the self's experience in relation to the world, so a radical distinction between emotion and reason in religion falsifies the integrity of religious experience. The history of theological reflection and philosophies of religion are replete with theories of religion that emphasize either the emotional or the rational nature of religion. And yet there are of course also those who work against disentangling religions' emotional and rational aspects. Gustafson cites both Augustine and Edwards as theologians whose understandings of religion account for the interpenetration of both. Gustafson is following in the line of these thinkers when he writes, "Our affectivity is evoked by our perception and understanding of others, of things objective to ourselves, as well as by subjective conditions."[35] The objective and subjective poles of religious affectivity here affirmed require that any account of the phenomenon of religion integrate rather than rigidly separate emotion and reason. To the extent that Gustafson's account of religion as affectivity accomplishes this, he claims that it avoids the historically debilitating tension between religious reason and emotion.

---

34. Though his assertion here is clear, its support is not very well clarified. What is implied, I believe, is that insofar as the affections are always related to some object, then their emotional dimension always includes a cognitive aspect. For to be affected by some object entails cognitive perception of it.

35. Gustafson, *Ethics from a Theocentric Perspective*, 1:200.

In addition to avoiding this tension, Gustafson argues that his account of religion circumvents the traditional theological and philosophical tension between faith and reason. His view of religion as principally affective means for him that "piety" rather than "faith" more adequately designates the religious sensibility. Piety for Gustafson is not "sanctimoniousness" but a settled religious disposition toward the ultimate object or objects in reality.[36] Whereas faith is traditionally contrasted with reason, piety based in affectivity is interrelated with reason. With a traditional emphasis on faith typically comes a contrast with reason. This contrast produces a distinction between and the necessity of balancing the relative epistemic significance of, on one hand, specifically religious sources of authority, for example scripture and tradition, and on the other more public accounts of reason and experience. For Gustafson, the contrast between faith and reason and the problem of authority it produces falsifies the complex inter-tangled density of human experience of ultimate reality. Piety as a settled disposition or attitude toward reality avoids this, according to Gustafson, insofar as it is evoked by experiences whose nature includes subjective senses and emotions, cognitive elements, and an objective referent. Having analyzed Gustafson's critical religious naturalism, it is now possible to consider how it shapes his view of the specific nature and tasks of theology.

Theology

As has already been indicated, theology for Gustafson is second-order reflection on religious experience. It is not a strictly theoretical exercise because it includes, is generated by, and responds to what he has characterized as the affective nature of religious experience. Theology is not only concerned with the arrangement of epistemological principles and metaphysical ideas. But neither is it only practical, for it includes theoretical principles, concepts and modes of argument. Though not only practical, Gustafson's view is that theology is principally an enterprise of practical reason. This derives in part from the priority he grants to experience. The theological enterprise is generated by the impulse to make sense of the multiple dimensions of human experiences, "to find some meaning in them and for them that enables persons to live and to

---

36. Ibid., 201.

act in coherent ways."[37] As not only but principally practical, theology is to some degree testable with reference to its consequences for the practical lives of those who are shaped by it and by the adequacy of its construals of the meanings of human experience.

Given this basic position, Gustafson affirms Julian Hartt's understanding that theology is fundamentally a way of "construing the world."[38] There are of course multiple ways of construing the world, secular and scientific in addition to religious, but each of these functions in similar ways to describe, explain, and evaluate the world. Distinctively theological world-construals refer understanding of all things in terms of their relation to God. Non-theological world-construals reference understanding to some other primary object, for example, the material or economic conditions of existence. But theological construals of the world also of course vary enormously among themselves. Theology is and has always been pluralistic and developmental. Given this, different theological construals of the world engender different ethical visions.

Differences derive from critical judgments about which theological tenets, symbols, or themes are most central to an adequate construal of the world, how these concepts are to be arranged or ordered, and these judgments together determine distinctive accounts of the moral life before God. "The dominance of certain theological tenets shapes the prescriptive ethics," writes Gustafson, "and deeply informs the analysis of the human predicament and the circumstances of human action."[39] Given the centripetal practical force of certain tenets, themes, and symbols, and Gustafson's emphasis on experience, ethical critique of theological world-construals must attend to the moral adequacy of these concepts to human experience and their place within the larger conversation of the ongoing development of theological traditions.

The hermeneutical moral question of what is going on in the world, and the specifically religious question of what God is enabling and requiring us to be and do, entail for Gustafson an evaluative description of experiences and their circumstances. Every moral act is initiated by an interpretation of circumstances that in lesser or greater degrees of consciousness is grounded in beliefs or convictions about

---

37. Ibid., 158.
38. See Hartt, "Encounter and Inference in Our Awareness of God," 51–54.
39. Gustafson, *Ethics from a Theocentric Perspective*, 1:161

the world. Such convictions, borne from pre-reflective experience, are always symbolically conceptualized.

The interpretive work of symbolic conceptualizations varies according to the particular beliefs they intend to conceptualize and to the ways in which those beliefs are held, whether strongly or weakly. Symbols are, for Gustafson, heuristic devices intrinsic to the evaluative description of circumstances. Particularly religious symbols control theological ethical interpretations of the circumstances of any moral experience. But the manner in which symbols operate and selectively determining which symbols should be central to interpretation and viewed as salient to the given circumstances vary. Individuals, as common sense well testifies, interpret events in multiple ways. So do theological ethicists, in spite of their possible shared commitments to a common tradition. This fact of variety, between theological and non-theological interpreters, among the non-religious and among the religious, raises at least two interrelated critical issues concerning the justification and criteria for the selection of certain religious symbols over others.

The justification of symbols of any kind hinges, in part, on the question of the epistemic authority of tradition. Gustafson attempts to locate his position on authority between two extremes: on the one hand, a confessionalist stance in relation to the Bible and Christian tradition, on the other, a tradition-indifferent ecumenical theological epistemology. In many accounts of theological ethics, the justification of specifically Christian symbols over others derives from some account of the autonomous authority of the Christian Bible. The symbols and themes contained within the Bible are automatically and intrinsically justified. Biblical symbols are justified as such on account of the distinctive ontological status of the Bible.

Gustafson treats the Bible's authority differently. The Bible cannot be treated as if it were uninfluenced by history, immune from interpretation, testing, or criteria from other domains of knowledge. The authority of the Bible is based for Gustafson on his conviction that it contains a meaningful human record of God's historical self-disclosures. The Bible and its symbols have authority, and their use in the interpretation of circumstances is justified, on the view that they disclose something about what is ultimately significant for human experience in the world before God. The Bible's symbols "disclose dimensions of events, of circumstances, and of persons that might not be perceived by other beliefs, other symbols, and other concepts."[40] Thus, the justification for the use of biblical symbols according to this line of reasoning does not depend on the peculiar ontological status of the Bible as such, but from experientially grounded subjective resonance with what others have experienced as ultimately and objectively real and significant.

With respect specifically to Christian theological convictions, the legitimacy of particular beliefs and the particular symbols conceptualizing them of course vary depending on one's relation to one among the many historical and developing traditions within Christianity. But to some degree or another, Gustafson writes, every tradition and community within Christianity by its very nature "claims that its symbols and concepts . . . have universal significance."[41] Though this universality may be a core claim of Christianity, and the projection of universal significance may be epistemologically unavoidable as part of the very meaning of holding to beliefs of any kind, it remains problematic. For it is an empirical fact that others outside of Christianity do not assent to the adequacy or significance of Christian symbols for their experiences. The question this forces is whether Christian theological construals of the world can be persuasive to others, aside from the point that for Christians these construals have universal significance.

Gustafson answers this question with a tentative affirmative. In support of this he affirms "a range of authority" between the extremes of a fundamentalist-literalist account of biblical authority and the view that authority is always and only socially constituted.[42] This affirmation rests in part on the assumptions cited at the beginning of this chapter—the ontological assumption that reality is unified, the epistemological assumption that various modes of inquiry can access different aspects of this reality, and the theological assumption that the experience of God in reality is multidimensional. The multidimensional nature of human experiences of God means that Christian symbolic construals of that experience are multi-referential. Experience is always of some or several dimensions of God in reality *and* something additional. Gustafson writes, "Religious persons experience the reality of God in and through

---

40. Gustafson, *Can Ethics be Christian?* 123.
41. Ibid., 128.
42. Ibid., 128–29.

experiences that at the same time are experiences of something else."[43] Every Christian symbolic construal of experience is multi-referential, acknowledging the specifically Christian theological significance of experience before God and the more general experiences of being in the world in relation to powers beyond human control and of the six common senses earlier discussed.

So the justification for the use of Christian symbols in the task of interpreting experience depends not only on the criteria internal to the Christian tradition but also on their interpretive leverage on the meaning of human experience generally. On this point Gustafson claims, "If a religious or theological interpretation of the significance of events does not render an account that is at least intelligible (if not convincing) to persons outside the religious circle, it obviously needs to be reconsidered."[44] Having said this, however, he is quick to make the point that reconsideration does not necessarily require capitulation. For though the significance of Christian symbols depends in part on their intelligibility to religious and non-religious outsiders, as with any other major religious tradition, they carry the weight of confirmation through the long history of Christian expressions of experience. This history bears its own considerable justificatory significance. For Gustafson, then, justification derives from the experientially verified authority of religious symbols as construals of the significance of events and experiences in the world. While criteria internal to the religious tradition matter in the question of justification, these cannot be the sole criteria. Human experience in general must also be appealed to.

Granted this account of the justification of symbols, the task remains of more carefully specifying the criteria for symbol selection. The importance of this should be obvious. Theologians often deploy different symbols in their interpretation of the significance of the same events and circumstances. Evidence for this is plentiful and illustrations need not be given here. So, criteria for symbol selection vary. But which criteria should control symbolic construals of circumstances and events? Gustafson stakes a position that is critical and complex.

The appropriateness of symbols to circumstances is the general concern, and the specific concern is to provide a theological defense

---

43. Ibid., 129.
44. Ibid., 129–30.

of this appropriateness. Shaping Gustafson's position is his fundamental position that the experience of God is multifaceted and that any account of such complex experience must therefore bring to bear numerous interpretive resources. Considering the complex nature of Christian religious experience, the various dimensions of the experience of God sometimes seem to conflict with one another. For example, the experience of God as judge is sometimes hard to correlate with the experience of God as redeemer, and the experience of God as ultimate power challenges our experience as moral agents with some capacity freely to determine and direct our own lives. In the effort then to select symbols appropriate to an interpretation of circumstances, judgments must be made about which general theme or dimension of experience ought to control the others.

For Gustafson, the experience of God as beneficent is the controlling criterion. But it is essential to understand here that Gustafson does not hold that divine beneficence always conforms to human conceptions of the good.[45] In Christian theological expression, Gustafson interprets his position on divine beneficence as parallel to the convictions that "grace is prior to nature," that "gospel is prior to law," and that "forgiveness is prior to sin."[46] These expressions each signify what Gustafson takes to be the basic intention of the divine, ultimately to seek the good of creation, even when that good may conflict with human conceptions of it. This position is certainly open to debate and as of yet Gustafson's theological defense of it has not been considered. That will be taken up in chapter four. For now it remains to examine how this criterion works in the question of symbol selection and construal.

While divine beneficence is Gustafson's controlling and first criterion for symbol selection, he articulates several additional specific tests: interpretive-critical, ethical, and social. The symbol needs to be appropriate to the circumstances. Determining this appropriateness is an evaluative exercise that requires interpretive self-reflexivity. The interpreter must critically engage the questions, for example, of the degree to which prior moral commitments or biases may be shaping symbol

---

45. This is, of course, precisely the issue at stake in my contention that Gustafson's motivating problem concerns the theological and ethical dilemmas of thinking and speaking about God in light of the—from finite human perspective—sometimes creative, sometimes destructive power of God in nature and history.

46. Ibid., 139.

selection. In addition, the symbols must be congruent with the ethical concepts of the religious community. The selection of symbols has to hold up to the scrutiny of others who are engaged in the same process of interpreting circumstances. Symbol selection needs to be defended within a community of interpretation.

Having articulated Gustafson's view of the criteria for theological symbol selection, an enterprise he deems necessary to conceptualizing the moral significance of circumstances and experience, it is important now to return to the issue of the various dimensions of experience. His arguments for the justification of religious symbols and the criteria of symbol selection support a significant claim. Theological interpretations, though arising from within specific religious communities, can disclose the general moral significance of common human experience and circumstances. In other words, what a theological interpretation of circumstances discloses need not nor should it be utterly unique. The significant content of experience is not tradition-specific, though the form of its construal often is. In fact, for a tradition-specific interpretation of circumstances to be significant it must at the very least be intelligible, if not persuasive, to those outside of the tradition. This point prepares the way for a discussion of Gustafson's account of the relation between theology and the sciences.

For Gustafson, as discussed above, the significance of any theological world-construal stems in large measure from its capacity to disclose the moral meaning of circumstances and the multiple dimensions of common human experience to a broad audience. One implication of this is that theological construals of the world, while aiming to be consistent with the tradition from which they are generated, can helpfully be informed by the sciences. For the sciences endeavor objectively to describe the world of nature and human existence, to provide accurate accounts of the world as it is. Given this endeavor, theologies that do not enlist the sciences in their construals risk neglecting reasoned, tested, and commonly held basic descriptions of the world. The ultimate theological loss wagered on this neglect is the possibility that a theological construal of the world, compared to a scientific interpretation, may become either an insignificant or meaningless disclosure of experience and circumstance, and thus need to be rejected. This is a potential loss that for Gustafson must be wagered according to his understanding of theology as an interpretive effort to disclose the meanings of common human experiences. Still, the question of how to relate theology and the sciences is a thorny one.

This question has received significant attention especially in recent decades. This is due in large measure to the many successful scientific contributions to our understandings of the world. These successes have yielded enormous confidence in scientific modes of reasoning. This confidence has also brought with it a moral optimism that the applications of scientific knowledge can provide strategies for living around if not eliminating the sickness, suffering, and poverty, among other contingencies, that have plagued humans throughout our history. This optimism sometimes translates into a normative scientism—the idea that scientific methodologies alone, apart from the other domains of humanistic knowledge, for example the arts and religions, warrant our ultimate allegiance. This of course poses a problem for religions, to the extent that religions claim generally to be about what is ultimate significant. In response to this perceived conflict between the sciences and the religions, a popular cultural view has developed that considers the relation between theology and the sciences antagonistically.

But the antagonistic model of the relation between the sciences and religions, apart from the question of whether it provides an accurate description of the history of their relationship, is only one among several ways of understanding how the sciences and theology interact.[47] Gustafson articulates his view of the proper nature of this interaction in contrast to several other models. First, he rejects the model that treats theological propositions as scientific ones, exemplified by some theologians and Christian believers who treat the account of creation in Genesis as literal truth. Second and conversely, he rejects the reduction of theology to the sciences. Theological reduction results when theological propositions are deduced from scientific ones. For Gustafson, theological and scientific modes of inquiry are different enough, and concerned with different dimensions of reality, that reductions from either direction are doomed to failure. Gustafson's judgment about this destiny to failure turns in part on his understanding that the sciences themselves are developmental traditions. Basing theological under-

---

47. One of the most influential theorists of the relation between the sciences and religion is Ian Barbour. In his work he describes four basic models of relation: conflict, independence, dialogue, and integration. On this, see vol. 1 of his 1989–1991 Gifford Lectures, *Religion in an Age of Science*.

standings on any historical period's scientific claims therefore freezes theological validity within a particular historical moment—a moment that the sciences have attempted accurately to describe, yet a description that may eventually be challenged by later scientific endeavor.

A third alternative against which Gustafson contrasts his model is based on an epistemology of "moderate realism" situated between positivism and idealism.[48] On this view, theological inferences are drawn from and support a scientific conception of the world as rationally ordered and of the human mind as capable of conceiving and explaining this order. The infinite can be inferred from the finite. Gustafson's critique of this model is based on his inherent epistemological skepticism. There is simply too much evidence, in his judgment, of social and historical variation about the meaning of reality to warrant the correspondence theory of truth upon which he claims realism tends to depend.

In contrast to these models, Gustafson articulates his account of the proper relation between theology and the sciences through a qualified affirmation of the views of German theologian, historian, and philosopher Ernst Troeltsch. For Troeltsch, though the idea of God is only accessible by way of religious belief, "'[religious belief] asserts a substantial content which must stand in harmony with other forms of scientific knowledge and also be in some way indicated by these.'"[49] Gustafson qualifies Troeltsch in two ways. First, he does not think theology and the sciences must or can always be in perfect harmony. Yet, the substantive content of theology "cannot be in sharp incongruity" with scientific knowledge.[50] Second, it is not so much religious belief as religious piety that ought not to be incongruous with the sciences. In other words, the congruity sought is between the sciences and the affective embodiment of the senses, attitudes, and dispositions of a religious sensibility. Gustafson's emphasis on piety rather than belief relates to the methodological priority he grants to experience, the affections, and the universal human senses.

---

48. Gustafson considers as an example of this Stanley L. Jaki's position in his Gifford Lectures, *The Road of Science and the Ways to God*.

49. Troeltsch, "Religion and the Science of Religion," 117; quoted in Gustafson, *Ethics from a Theocentric Perspective*, 1:251.

50. Gustafson, *Ethics from a Theocentric Perspective*, 1:251 (italics mine).

Gustafson's theory of the proper relation between theology and the sciences is, then, characterized by several interlocking claims. First, theology and the sciences are relatively autonomous. They are not incommensurable or non-overlapping, but neither are they perfectly symmetrical or always harmonious. Theological propositions should not be treated as scientific claims, nor should the claims of theology be entirely limited to what can be deduced from scientific theories. Though neither incommensurable nor perfectly harmonious, if theology claims to be about the reality of human life in the world before God, "evidences from the operations of nature [should] be taken into account . . . ."[51] Second, and given the above, if scientific knowledge about the world must necessarily be included in the tasks of theology, then the validity, intelligibility, and plausibility of theological world-construals depend in part on testing them in relation to scientific theories and data. But again, the sciences do not provide the sole criteria for theological validity, for the sciences and theology are relatively autonomous, concerned with simultaneously different and similar orders of experience in the world. Third, there is no such thing as "science" in the singular. One can only speak of the "sciences." Different ranges of explanation and interpretation are correlated to different scientific fields that are under nearly continual revision, their objects of study, and the methodologies appropriate to those objects.

Given these three basic commitments, several questions arise. These include, as Gustafson articulates them:

> What indications from the sciences are relevant to what is said about the characteristics of the powers that bear down upon us and sustain us? What data and theories from the sciences are appropriate tests of the intelligibility and plausibility of theological affirmations? Which are proper tests of a theological construing of the world? With which theories must theological statements be congruous?[52]

Not all theologies will negotiate these questions in the same way, but any theology must consider these questions if it accepts Gustafson's model of theology and the sciences. As will be shown in chapter four, Gustafson's own construal of the world and humans turns on a quali-

---

51. Ibid., 258.
52. Ibid.

fied Reformed theology. This theology will require that these questions be answered in distinctive ways and will yield a specific construal of the congruity between Reformed piety and scientific insights. However, before moving on to that work, and to the treatment of Jonas's fundamental ideas in chapter three, I will conclude with an assessment of where we have come thus far.

Conclusion

The preceding interpretation of the hermeneutical dimensions of Gustafson's work, the moral motivation backing it and the methodological commitments informing it, suggests some significant points of contrast with Jonas. Examining these will advance the endeavor of delineating the role of their differing moral anthropologies within their larger projects and in relation to my own purposes.

As explained previously, Jonas's driving concern is with the environmental ethical implications of the moral ambiguity of human power. The environmental crisis is a symptom of this and an occasion for Jonas to provide a biologically informed philosophical reconstruction of the human place within the phenomenon of life. This reconstruction is undertaken in order to develop an anthropology and ethics of responsible human power, to be explored in chapter five. The opening of this chapter characterized Gustafson's primary concern as that of human moral evaluations and theological conceptions of divine power as these inform thinking about God and the human good. Whereas Jonas is repulsed by the horrible potential of human power's ambiguity, Gustafson stands in awe before a power beyond human control, a power that he perceives as simultaneously beautiful and horrifying, as sustaining and devastating, and thus a power that poses conceptual and practical problems concerning theological ethical discourse and religious moral life.

Thus, while both Jonas's and Gustafson's projects are morally motivated by a concern with power, the horizons of their concern are significantly different. Gustafson's focus is not on human power but more expansively on the powers of nature and history, ultimately divinely governed. His horizon of concern is therefore more comprehensive and more complicated than that of Jonas's. I have also indicated that Gustafson's project is not as explicitly motivated by the environmental crisis as Jonas's. Instead, Gustafson concentrates on the theological and ethical incongruity of anthropocentrism in light of the human experience of life's dependence on a power beyond human control.

While Gustafson's writing and thinking are theological, his theology is methodologically structured by a critical religious naturalism built upon a phenomenology of the basic contours of common human experience. Through this approach, he argues for the multidimensionality of human experience and for the six common human senses of being in relation to powers or a power that bears down upon and sustains life. These senses need not, but can be construed religiously if they are understood in relation to ultimacy. In this way, and in contrast to Jonas's subtle emphasis on organismic autonomy, Gustafson's methodology reflects a concern with the interdependent character of human experience, leading him to emphasize the finite limitations of human power, purpose, and freedom.

Gustafson's crucial insight on this score, and in contrast to Jonas, is that purposiveness is not fundamental but derivative. While freedom and purposiveness may be common traits of organisms, as Jonas argues, they are, according to Gustafson, constituted through, situated within, and limited by the patterns and processes of the larger environment. Gustafson's vision of nature, then, includes an emphasis on the relational patterns and processes of nature that sustain life, which, for Jonas, are peripheral to the more basic agonistic struggle between organism and environment.

In the conclusion to the previous chapter, I suggested that Jonas's critical existentialism, while extended to the whole of the organic world rather than to humans alone, harbors an idiocentric or individualistic residue within it. His approach brings helpfully to view the purposive character of life in general. This allows him, in his development of his moral anthropology, to stress the importance of human intentionality, and, in his ethics, of the need for this to be normatively guided by the counsel of responsibility. And yet, at the same time that he critiques Heidegger for recapitulating the historic divide between the human and the natural, Jonas's own methodology can be critiqued for yielding a dichotomy between the organic and the inorganic. This dichotomy reflects an atomistic versus a relational interpretation of life and life's evolutionary conditions. He focuses on a part of nature, purposive individual organismic striving, expressed primitively through metabolism, at the expense of a sense for the broader context that underlies, enframes, and provides the conditions of possibility for this striving.

Gustafson's methodology achieves a broader vision of the context of life that is subtly neglected by Jonas's methodology, and elicits a vision of life's dependence and interdependence. Further, Gustafson's understanding of the relational dynamic between life and life's contextual conditions shapes and is shaped by an account of moral experience that is more variegated than Jonas's. While moral experience for Gustafson always includes the intellect, will, and feeling, any unipolar treatment of such experience is reductive. The moral life cannot be described adequately in strictly cognitive, voluntaristic, or emotive terms because it always includes each of these dimensions. This multidimensional account of moral experience situates the human in a different position in the continuum with nature. As suggested above, though Jonas attempts to overcome the dualism between humans and nature, the accent on the purposive will reflected in his methodological commitment to existentialism can be interpreted as producing a dualism between the organic and inorganic. Continuity between humans and other biotic forms is affirmed through this methodology, but this affirmation rests, again, upon a subtle discontinuity between bios and the material environment.

If "nature" includes more than the living, and surely it does, then the position of Jonas's emerging view of moral anthropology on the continuum with nature leans toward discontinuity. The human is related to other forms of life in a graded scale of complexity, but as with all other forms of life, according to Jonas, exists in an agonistic relationship with the broader context of the natural world. In isolating purposiveness as the essence of the organic, Jonas's philosophical reconstruction of nature reflects a breech between bios and the environment.

Gustafson's critical religious naturalism and the density of his account of moral experience place the human in more significant continuity with the larger natural world. The human, like all other forms of life, is dependent on material natural conditions beyond our control. Certainly organisms shape their material environments, and human life has the capacity to do this to a radical degree, but however life alters the material environment, the environment always works back upon life. Life may be purposive, as Jonas argues, but according to Gustafson it is dependently embedded in a broader reality that always at least partly escapes the capacity of life to manipulate it. Purposiveness, in other words, is dependent on sustaining relations of interdependence. Thus, Gustafson's methodology and the view of moral experience it discloses show that he recognizes to a degree that Jonas does not the relational ties that bind the organic to broader historical, natural, and ultimately divine processes.

An additional very prominent difference from Jonas is that Gustafson's project attends very carefully to religious experience and the ways in which it can be theologically interpreted. The multidimensional character of human experience universally includes within it the common senses of dependence, gratitude, obligation, remorse, possibility, and direction. These senses, on Gustafson's interpretation, provide the basis for characterizing religion as natural. This is the most significant contrast between Jonas and Gustafson yet considered. As previously indicated and as further chapters will detail, Jonas views theology as peripheral to his project. For Gustafson, in pronounced contrast, theology is integrated in each of the dimensions of his ethical theory. The differing functions of theology in Jonas's and Gustafson's work have interpretive, descriptive, and normative ramifications. This chapter has considered Gustafson's view of the interpretive significance of theology. In chapter four on the fundamental ideas in Gustafson's project, the descriptive significance of theology will be traced. And this of course leads to the work on the thinkers' normative proposals. It is in their normative programs that the rubber of interpretation and description hits the practical road.

Already, however, a pivotal normative difference stemming from their moral anthropologies has been foreshadowed. As discussed earlier in this chapter, Gustafson's critique of our circumstances adds a concern with the inwardly curved nature of the human condition to Jonas's concern with the morally ambiguous effects of human power. This critical supplement of Gustafson's reflects a theological insight that Jonas lacks. For Jonas, as will be shown in chapter five, the remedy to the emergency nature of the environmental crisis is to root morality in the nature of things, rationally to demonstrate the objective ground of value and obligation. The assumption is that such a demonstration will be sufficient to a responsible redirection of irresponsible, shortsighted human behavior and policy. He argues that what is worth preserving is intuitively obvious. In other words, what the future of humanity and the rest of life demands is the rational direction of human power by a proper understanding of the intuitive value of nature.

Gustafson's position on this is very different. The moral problem for Gustafson is not principally one of reason but of the affections. The will is not moved by reason alone. Gustafson's constructive task, given his focus on the distortion or contraction of human affections, is not rationally to demonstrate the objective or intuitively obvious foundations of morals. For this would not change much, given the human proclivity toward narrow self-concern. His aim, rather, is persuasively to point toward the expansive life of piety toward which the natural senses of the world can be interpreted to correspond. The contraction of the human spirit yields a false sense of the significance of our species. Against this contraction, the life of piety Gustafson calls for will be constituted by the dilation of our feelings, attitudes, and disposition toward the singular power of God that works through the myriad of powers that bear down upon and sustain life.

But this is getting slightly ahead of things. Suffice it to say, at least, this chapter's interpretation of Gustafson's methodology and the moral motivation driving his work begin to bring into view the normative significance of a distinctive moral anthropology. And yet, while different, I will continue to make the case that each thinker's anthropology includes aspects that are complementary rather than opposed, and that if joined together present a more fully dialectical anthropology that may hold promise for a time of environmental crisis.

PART TWO

Fundamental Dimensions

3

Technos, Bios, Anthropos

Introduction

As with the preceding interpretations of the hermeneutical aspects of Jonas's and Gustafson's work, two aims guide my next two chapters: to exegete crucial dimensions of each thinker's broad project, and to gather from these exegeses each thinker's insights for an ecotheological ethics of responsible participation. The dialectical rhythm of this book's rhetorical structure, moving from Jonas to Gustafson and back, reflects my constructive purpose. An ethical-religious vision adequate to the moral demands of our time and to the human relation to nature, I am arguing, needs to be dialectical. That is, it needs to do justice to both the continuity of humanity within the natural world and to the gravity of the specifically human moral burdens we bear within it. A critical interpretation of Jonas's fundamental ideas moves this project further toward this vision.

Jonas's effort to think with and beyond the categories of his mentor Heidegger, traced in chapter one, provides an important background for this chapter's account of his basic descriptions of the world, life, and human beings. For the moral urgency of these descriptions is brought into view under the light of the critique of dualism previously engaged. Against the first chapter's interpretation of his hermeneutical horizon, Jonas's fundamental descriptions can be seen as the platform for the treatment of his normative proposal in chapter five. Thus this chapter serves as a pivot between Jonas's methodological and prescriptive tasks.

Though I have divided the fundamental dimension of Jonas's theory into his separate ideas about the world, life, and human being, it is important to remember, and I will attempt to show this, that these ideas intermingle. While referring to different aspects of reality for Jonas, technos, bios, and anthropos are united by a central theme—the descriptively and morally profound connection between *being* and *doing*. The being of all living forms, but emphatically so for human beings, is constituted according to Jonas as activity. The efficacious range of this activity varies among the forms of life, but at its origin is essentially purposive, and this native purposiveness serves for Jonas as the principle for the reunion of being and value.

Technos

The emergence of the world as technos, or as shaped by a technological ethos, follows for Jonas from select historical conceptual and moral revolutions. The first revolution I will interpret concerns the dialectic between Athens and Jerusalem, or philosophy and religion. I will examine here Jonas's idea that the doctrine of creation is a crucial factor in the technological ethos, insofar as it ironically results in the moral diminution of the natural world and the moral elevation of humanity. In so doing, this revolution paves the way for the second revolutionary moment to be considered: the conceptual shifts that, on Jonas's interpretation, shape the character of modern science. One of the general effects of the scientific revolution, on Jonas's interpretation, is the abstraction of the world that this revolution simultaneously depends upon and reinforces. Following my interpretation of these theological and scientific conceptual revolutions, I will examine their connection to what Jonas takes to be the precarious situation of our technological age—the applied results of the elevation of the human and the world's diminution and abstraction in the absence of a durable ethics.

Historical Background

Jonas's morally charged study of the parallels between Gnosticism and existentialism discloses his view that ethics and metaphysics mutually shape each other, even when this is explicitly disavowed. Jonas's position here critically runs against modern ethical theory's tendency to abandon metaphysics. Against this tendency, Jonas emphasizes the need for ethics and metaphysics to be reunited after their modern eclipse. Ethics must be tutored by metaphysics in order to be sufficiently grounded and to provide motivating direction for the morally responsible life under the radical pressures of contemporary circumstances. But before an adequate treatment of this reunion is possible, it is important to consider further Jonas's understanding of their divorce.

In "Jewish and Christian Elements in Philosophy," Jonas provides a penetrating account of the influences of Jewish and Christian thinking on the character of Western philosophy and modern scientific sensibilities.[1] According to Jonas, of these many influences, the Jewish and Christian idea of the world as created is the cornerstone of this heritage.

The central philosophical contribution of a theology of creation is the conception of the world as *historical*, as having a beginning in time contingent on divine will. This is diametrically opposed to the classical notion of the world's timeless co-eternity with a supramundane principle. Uniting various classical philosophies of nature is the idea that the world is a necessary extension of divine nature. On this view, given the eternity of the divine and the emanation of the world from God, the world too is eternal. Because the world's essence is eternal, so is its order. The world cannot be other than it is. The necessity of the world being as it is and not otherwise derives in classical thought, according to Jonas, from the dominant view that the world emanates from an intellectualist conception of God's essence. The divine nature, the locus of necessity, is principally rational. Thus, as Jonas argues, the principle of the world's eternity is bound up with a metaphysics from which that principle derives.

Likewise, the theological view of the world's temporality is characterized by a number of key subversions of classical concepts. With the idea of "creation," Jonas argues, conceptual priority is granted to contingency over necessity, particularity over universality, and divine will over reason. These conceptual adjustments, as will be shown, have enormous practical as well as theoretical significance for the idea of nature and the human relation to it.

This significance turns most specifically on the biblical emphasis on the world's historic contingency, fundamental to the biblical account of divine omnipotence. The world as contingent rather than necessary, as historically dependent on its Creator, severs the classical identity of the eternity of the world and the divine. The world's diminution is

---

1. Jonas, *Philosophical Essays*, 21–45. Originally presented as Third Lentz Lecture at the Colloquium on Judaism and Christianity, Harvard Divinity School, October 1967.

proportionally and paradoxically correspondent to this emphasis on contingency, according to Jonas: "'createdness' implies a certain devaluation of that which has this attribute...."[2] The world is not made in the image of God but is rather an effect of divine craftsmanship. The *imago dei* is a status reserved for humanity as God's special creation. The biblical emphasis on the world's contingency ironically results, then, in the diminution of the world, and, on Jonas's score, this diminution underlies the modern scientific reification of nature.

According to Jonas, the "ex nihilo" tradition problematizes matters further. For the emphasis on divine power at its core, intended to signify and honor God as the transcendent source of all things, eventually and ironically leads to an emphasis on human power and will. Jonas writes that the importance of "ex nihilo" is the relationship between value, will and power that it establishes. So long as value and power are united in the wisdom of the divine, value and norms are secure, "[but] when that ground vanishes, as it does with the vanishing of faith, there is only *man's* will and power to ground any norm or law."[3] When the world's contingency, originally signifying divine power, becomes severed from divine origin and from a theological epistemology, the result is the eclipse between being and value that, for Jonas, characterizes the modern moral situation.

Through scientific challenges to the theological mind, the human replaces God as the inventive source and ground of value. As I will explore more fully further along, Jonas later reinterprets the idea of creation and the divine relation to the world and humanity in a way that, while building upon this replacement, morally radicalizes it in order to emphasize the profound human responsibility for the future of life. But this is getting ahead of things. In what follows, I turn to Jonas's interpretation of the principal conceptual revolutions that drive the history of modern science.

Jonas's treatment of modern science focuses on two principal conceptual shifts, cosmological-descriptive and physical-explanatory,

---

2. Ibid., 35.

3. Ibid., 43. This judgment about the "vanishing of faith" is of course debatable. But according to Jonas, and this is unquestionable, faith in a Creator god is challenged by the rise of scientific reasoning. Jonas's views on aspects of the "modern" questioning of religious faith will be considered in ch. 5, through interpretation of some of his theological writings.

which together contribute significantly to the dissolution of the ideas of natural hierarchy and teleology. He traces the first of these shifts to Copernicus's famous argument for a heliocentric understanding of the universe.

Jonas suggests that the Copernican turn contains three especially revolutionary ideas: cosmic homogenization replaces cosmic hierarchy, cosmic kinetics takes the place of cosmic architectonics, and the picture of the cosmos becomes infinitely expanded. First, cosmic homogenization results from the heliocentric redescription of the earth as one star or planet among others, not qualitatively distinct or nobler than the rest. This homogenization entails the collapse of the traditional hierarchical interpretation of the universe and the earth. Jonas identifies the second revisionary idea of Copernicus's heliocentrism in terms of the replacement of a cosmic architectonics by a cosmic kinetics. The pre-Copernican, Aristotelian-Ptolemaic cosmology explained the motion of the celestial spheres in terms of the axiom of circularity. Within this cosmology, the circular motion of the celestial spheres was understood as caused by the impulse of a First Mover exerted at the outermost sphere of the cosmos rather than by a dynamic of multiple interacting forces. Through a combination of a new explanatory mathematics, the leveling of celestial nobility resulting from cosmic homogenization, and Kepler's thesis of elliptical orbits, Copernican heliocentrism challenged the explanatory power of this architectonics and called for a new account of kinetic causality.

Cosmic homogenization and the collapse of cosmic architectonics lead, on Jonas's reading, to a third major conceptual revision, the infinite extension of the universe. To all appearances the earth does not move. But according to Copernicus, the stars that *appear* to move in relation to a supposedly unmoving earth only *seem* so to move because of their enormous distance from observers. Heliocentrism thereby, in Jonas's words, "monstrously magnifies" the universe to unimaginable dimensions. The earth is neither the center of the cosmos nor is there any reason to believe that the stars and planets we can see from our planet constitute the limits of the cosmos. Infinitely expanded, the universe has no center.

Cosmic homogenization, the dissolution of cosmic architectonics, and the infinite expansion of the universe certainly lead to a radically expanded world-picture and a radically diminished sense of the human place in the cosmos. United, according to Jonas, these three conceptual shifts challenged traditional notions of cosmic and natural hierarchy, and called for a new explanatory physics. The work of developing this new physics, according to Jonas, constitutes the second conceptual shift underlying modern science and leads to a mechanical conception of nature. Through this shift, the end of teleology is added to the Copernican challenge to cosmic hierarchy.

On Jonas's account, the second conceptual revolution within modern science is led by the demand for a new theory of motion required to explain the Copernican cosmology. This conceptual revolution is initiated by Galileo, and then completed by Newton. For Jonas, the explanatory function of abstraction is of crucial significance here. As with Copernican heliocentrism, Galilean kinetics "replaces the testimony of the senses with an abstraction that directly contradicts but indirectly grounds it."[4] Explanatory efficacy is demonstrated at the level of mathematical abstraction rather at the level of experiential perception, and this influences the rise of methodological reductionism and experimentalism. On the Galilean view, change of motion is explained "downward" through reduction to its composite forces rather than "upward" toward the final cause of the First Mover. Explanation of motion hereby takes on a reductive caste. Further, the reductive analysis of the forces constituting kinetic change "permitted an actual dissociation of [the] component parts [of motional change] in suitably set up *experiments*: it thus inspired an entirely new method of discovery and verification, the experimental method."[5] One of the important consequences of this development, for Jonas, is that nature comes to be conceived in the abstract, and thus more open to manipulation.

On Jonas's reading, Newton's mechanics completes Galileo's kinetics in this second stage of the scientific revolution. Newtonian mechanics centers on the definition of the concept of "force" so important to the new cosmology and its revisionary understanding of motional change. Galileo had posited that all change of motion could be described in terms of velocity, the rate of an object's change of position, or acceleration, the rate of change of velocity. All change of motion can be represented as different values of these concepts. But still, the ques-

---

4. Ibid., 61.
5. Ibid., 63.

tion remains as to what operative force is at work in either velocity or acceleration. In other words, what is it that pushes or pulls to change a body's state of motion?

Newton answers this with his "Universal Law of Gravity," through which he explains that the same force pulling objects downward on earth operates as well for the moons and planets beyond the earth. Any two objects are attracted to each other according to the exertion of a gravitational force between them. The strength of this attraction is proportional to the product of the objects' gravitational masses, and inversely to the square of the distance between them. Newton of course did not "discover" gravity. What Newton contributed to scientific understanding was an extension of the laws of nature beyond the earth, which, contrary to the previous Aristotelian-Ptolemaic cosmology, was not a phenomenally distinctive realm. In this sense, Newton's Universal Law of Gravitation completes the conceptual trajectory initiated by Copernicus by specifying the cosmic continuity of nature's laws.

In sum, the revolution initiated by Copernicus and fulfilled in at least one of its dimensions by Newton leads to a view of nature as a fully explicable mathematical abstraction, understood through reduction, replicable through experimentation, and open to manipulation. Jonas's view of the metaphysical consequences of these conceptual shifts has metaphysical consequences that provide the springboard for the philosophical reinterpretation of nature at the core of his ethics.

Of primary metaphysical significance here is the impact the new cosmology and physics have on the idea of causality. Whereas in the pre-modern Aristotelian-Ptolemaic metaphysics, final causation held primary explanatory power, in the new cosmology and physics, this emphasis is replaced by efficient causality. This new emphasis challenges a teleological, purpose-directed vision of nature. Neither supramamundane divine causality nor intramundane human mental causality is afforded room in the explanatory regime of the new science. That is to say, in a strictly mechanistic theory of change, purposiveness cannot figure as a causally explanatory concept. Rather, causality comes to be understood through the concepts of force, velocity, acceleration, and gravity inherent to the understanding of matter produced by the Copernican homogenization of the cosmos. The efficient power of mathematically calculable force replaces purposiveness in the theory of causality. All change in nature hereby becomes quantifiable and predictable according to mathematical abstraction.

The metaphysical fallout of this is that for the descriptive and predictive capacity of mathematics to hold, supramaundane intervention into the processes of nature has to be ruled out. This is a metaphysical corollary of the new cosmology and physics. In short, radical determinism goes hand-in-hand with the new science's mechanistic reduction of change to efficient causality.

According to Jonas, this has anthropological as well as metaphysical and theological significance. For if the new idea of causality accounts for change in terms of strictly material antecedents, ruling out supramundane intervention, it also rules out "the most ordinary initiation of an external change by an act of *human* [will], no less than by one of divine . . . ."[6] This, of course, is an offense to probably the most basic of human intuitions, that humans are causal, intentional, purposive agents, seeking outcomes and impacting the world. But such an understanding is theoretically compromised, since it "would start a new causal train" not reducible to the determined material mechanisms of the laws governing the physical world.

In this way, to the end of hierarchy entailed by the Copernican description of the heliocentric universe, the new physics developed to explain that universe compromises the explanatory relevance of purposiveness. As the Copernican cosmo-descriptive revolution dismantles cosmic hierarchy, so the physico-explanatory revolution developed in response compromises any teleological interpretation of the universe. Jonas writes, "That Nature is devoid of even the most unconscious bias toward goals . . . that final and formal causes are struck from its inventory and only efficient causes left, follows simply from the principle of quantitative equivalence and invariance in cause-effect relations which is the distinguishing mark of the 'determinism' of modern science."[7]

To the gain of explanatory neatness provided by the new cosmic calculus, there are for Jonas profound moral losses. Foremost among these is the conception of nature as wholly indifferent to itself and to the forms of life produced and sustained by natural processes. "In the working of things," writes Jonas, "there are no better or worse results—

---

6. Ibid., 67.
7. Ibid., 67–68.

indeed there is no 'good' or 'bad' in nature, but only that which must be and therefore is."[8] This conclusion, to be sure, is adequate to Pascalian existential dread. The purposive, intending human is an anomaly within this world-picture. But in addition, and perhaps of even more consequence, a nature conceived as "indifferent to itself solicits no respect."[9]

A picture of nature devoid of purpose and soliciting no respect, indifferent to its own processes and the varied forms of life those processes generate, is deeply problematic when combined with a radical increase in human power through technological innovation. The gravity of this, of course, is the motivating source for much of Jonas's philosophy. In what follows, I will examine Jonas's assessment of the morally distinct character of the modern technological ethos, which, when interpreted against the background of an indifferent, purposeless, wholly accidental world of nature prompts the rigor and urgency of his phenomenology of life and moral anthropology.

An Ethical Novum

According to Jonas, one of the great ironies of the contemporary world is that lack of certainty about *how morally to be in the world* is in part a product of modern scientific advances in *knowledge about the world*. In other words, the contemporary situation is one in which humans have greater knowledge of and power over nature and yet less certainty about how morally to direct that power than ever before. This paucity of moral confidence is related to the anomaly of human existence created by the modern scientific worldview, in which purposive human experience is incommensurable with a vision of nature that is indifferent to itself and the forms of life it generates. The human is adrift in nature, riddled with existential anxiety. This anxiety is especially problematic morally in light of the technological extension of human power.

Jonas holds as axiomatic that responsibility is a correlate of power. Since his interpretation of the history of science and technology concludes that the magnitude of contemporary human power is of an unprecedented scale, he understands his philosophical burden to be that of providing a moral theory of responsible power objectively grounded

8. Ibid., 69.
9. Ibid., 70.

in the nature of things. Only such an objectively grounded theory of responsibility, he will argue, can be commensurate to the radically expanded nature of contemporary power. Such a project requires theorizing a philosophy of nature that can "bridge the alleged chasm between scientifically ascertainable 'is' and morally binding 'ought.'"[10] Theorizing this entails challenging the separation of fact and value, description and prescription, that is the legacy of the modern scientific picture of nature and one of the linchpins of modern ethical theory.

Urgency is added to difficulty since, on Jonas's view, the modern scientific view of nature and the novel character of modern human power combine to render all previous ethical systems inadequate. This marks a profound difference from Gustafson, for whom we must work critically to revise our existing historical moral and religious traditions. But for Jonas, the unprecedented moral situation of contemporary human power means that we cannot fall back on the resources of the history of ethics for guidance. A new ethics is required to meet the novel demands of our contemporary situation. It is not merely the case that science and technology have introduced new objects and problems of moral concern, though they have indeed done this. Rather, and more radically, "the qualitatively novel nature of certain of our actions has opened up a whole new dimension of ethical relevance [the future of life] for which there is no precedent in the standards and canons of traditional ethics."[11]

Jonas interprets several causal sources for this. First, and in contrast to the historical sense of nature's *invulnerability*, the consequential magnitude of modern technology reveals nature's fundamental *vulnerability*. Whereas human action historically has been limited with respect to its impact on nature, the immense power of contemporary technology is capable of unlimited and total impact. Second, and as a consequence of the above, whereas the history of ethics has been concerned with the relatively proximate sphere of intrahuman social action, the power of human technological action over nature magnifies the scope of moral concern. This expanded horizon of moral concern requires the temporal and spatial expansion of ethical thinking. Classical ethics

10. Jonas, *Imperative of Responsibility*, x.

11. Jonas, *Philosophical Essays*, 4. Originally presented as a plenary address to the International Congress of Learned Societies in the Field of Religion, Los Angeles, September 1972.

was concerned with a relatively "short-term context" correlated to the "short arm of human power."[12] Modern ethics, in contrast, must face the lengthening arm of human power and the expanded context of its effects. For Jonas, technology, rooted in the conceptual revolutions of modern science, introduces the whole of the biosphere and the destiny of humanity as moral objects, and this fact requires renewed ethical thinking.

At the Joseph Wunsch conference held in Haifa and Jerusalem in 1974, Jonas gathered with a group of major thinkers for an international conference on the theme of "Ethics in an Age of Pervasive Technology."[13] Several positions garnered unanimous support at the conference. These included the *magnitude and scale* of contemporary moral and policy issues. The moral problems generated by the pervasiveness of technology are not concentrated or isolable to small groups of people but are of enormous global consequence. In addition, the thinkers agreed that the *universal scope* and the *interconnectedness of effects* in a technological age constitute an unprecedented situation for ethics and policy. Through technology, morally relevant effects of action are not only immediate but, more significantly, future-oriented. The conferees also concurred that while most ethical traditions have been concerned to some degree with the *unintended consequences* of human moral action, contemporary technology significantly raises the stakes of this concern. For contemporary technology in many respects supercedes capacities for predicting and forecasting the outcomes of its uses.

In addition to these insights, the conference pondered the *exponential, logarithmic growth* of resource use and depletion and the *accelerated pace of change* as morally relevant factors endemic to our contemporary situation. Technology's effects are cumulative, enmeshing us all in rapidly changing artificial environments through the exploitation of natural resources. The positions that emerged form this conference permeate Jonas's thinking on the historical developments of science and technology.

In "Toward a Philosophy of Technology," an essay written for the *Hastings Center Report* in 1979, Jonas analyzes several of the distinctive

12. Ibid., 8. See also Jonas, *Imperative of Responsibility*, 6.

13. The proceedings of this conference are published as Kranzberg, ed., *Ethics in an Age of Pervasive Technology*.

formal characteristics of modern technology.[14] Most basically, Jonas characterizes modern technology as "an enterprise and process" in contrast to earlier understandings of technology as a "possession and a state."[15] Historically, the possession and state of technology as a limited inventory of tools and procedures was determined by an "equilibrium of ends and means."[16] Technologies were developed and utilized with reference to given ends of human desire and need. Though this equilibrium was sometimes altered, innovation was generally incremental, relatively inconspicuous, and linear. In contrast to this, the processes of modern technological innovation are totalizing, radical, and circular. As new technologies "suggest, create, even impose new ends, never before conceived," they radically transform "the very objectives of human desires."[17] In addition, the pace of innovation is quickened. This occurs in part as a result of the ease and efficiency of knowledge transmission, itself a product of the technological enterprise, and in part as the effect of the competitive market pressures in which technology tends to flourish.

Added to these points, Jonas interprets modern technological innovation as a "juggernaut." Innovation does not remain simply an option but is inevitable and guaranteed by the necessity of modern technology's cumulative-collective formal nature. This is of profound moral significance for Jonas. For it is in light of this that Jonas claims that the sphere of intrahuman "neighbor" ethics "is overshadowed by a growing realm of collective action where doer, deed, and effect are

14. Jonas, "Toward a Philosophy of Technology," 36. Jonas identifies several sources shaping the distinctive formal character of modern technology. Market pressures have already been mentioned. In addition to and as a further aspect of this, Jonas highlights the significance of population growth and the competition for finite resources needed to sustain this growth. Market competition and an expanding population are not only causally at the roots of technological process but are also effects of it. Medical technologies, for example, extend average life spans, reduce infant mortality rates, and thus contribute to population growth. With this in mind, Jonas writes that the market and population "offer a good example of the ... general truth that to a considerable extent technology itself begets the problems which it is then called upon to overcome by a new forward leap." Adapted from Jonas's presentation on the occasion of receiving the Hastings Institute second Henry Knowles Beecher Award for lifetime contributions to ethics and the life sciences, made at the Hastings General Meeting, June 23–24, 1978.

15. Ibid., 34.
16. Ibid.
17. Ibid., 35.

no longer the same as they were in the proximate sphere . . . ."[18] The actions of agents and the consequences of those actions in a technological age spiral in their moral relevance well beyond traditionally intra-human frameworks of historical ethical systems. Agents, actions, and consequences collectively "snowball" in their causal efficacy through the various forms of technology. Technology magnifies power through a cumulative expansion of the powers of individual and institutional actors. As a result, "an object of an entirely new order—no less than the whole biosphere of the planet—has been added to what we must be responsible for because of our power over it."[19]

So, in addition to the *collective* nature of technological action, human power today has a profoundly *cumulative* character. Jonas writes, "The containment of nearness and contemporaneity is gone, swept away by the spatial spread and time span of the cause-effect trains which technological practice sets afoot . . . ."[20] Consequences of actions pile up on one another, effects are added to effects and actors in the present significantly preshape the moral situations in which future actors find themselves.

Previous ethical systems, according to Jonas, tended to evaluate action on the assumption that it was noncumulative. Since human action without the aid of radical technological power was incapable of significantly altering the structure of the world, the relation between behavioral cause and effect was largely predictable. This is not the case in a technological age, for the "cumulative self-propagation of the technological change of the world constantly overtakes the conditions of its contributing acts and moves through none but unprecedented situations, for which the lessons of experience are powerless."[21]

The unprecedented scale of human power in a technological age and the unique moral situations this power introduces, lead, for Jonas, to making *knowledge* "a prime duty beyond anything claimed for it heretofore, and the knowledge must be commensurate with the causal scale of our action."[22] Though primary, the duty to knowledge created by

18. Jonas, *Imperative of Responsibility*, 6.
19. Ibid., 7.
20. Ibid.
21. Ibid.
22. Ibid., 7–8.

technology is also by the nature of contemporary technology doomed to failure. For contemporary technological practices outstrip our capacities for the knowledgeable prediction of consequences. Given this, Jonas claims that the "recognition of ignorance becomes the obverse of the duty to know and thus part of the ethics that must govern the evermore necessary self-policing of our outsized might."[23] This claim, that acknowledged ignorance must become morally governing, is central to Jonas's account of the moral significance of a "heuristics of fear," which I will interpret in ch. 5.

For now, it remains to consider Jonas's insight that in the ever-incomplete effort morally to direct power in our technological world, an important initial point of departure is to inquire into the moral nature of the species in which technology has its origin. Given Jonas's biological existentialism, this anthropological inquiry needs to be situated within a broad philosophy of life as such, and to this I now turn.

Bios

Jonas's sees his philosophical task in light of this diagnosis of our technological age as that of providing an ethical theory adequate to the new demands of human power. He writes that, "The new kinds and dimensions of action require a commensurate ethic of foresight and responsibility which is as novel as the eventualities that arise out of . . . the era of technology."[24] This is an ambitious aim. For it is not only the case that Jonas claims with his call for novelty that we have nowhere to turn in the history of ethics for contemporary guidance. It is also the case that the very technological and scientific advances that have generated the demand for moral direction also radically problematize it. Technological power demands moral direction and yet at the same time has eroded confidence in the foundations of moral norms. As Jonas puts this, "We need moral wisdom most when we believe in it least." Or, more graphically, "Now we shiver in the nakedness of a nihilism in which near-omnipotence is paired with near-emptiness, greatest capacity with knowing least for what ends to use it."[25]

---

23. Ibid., 8.
24. Ibid., 18.
25. Ibid., 21, 23.

The ambition of Jonas's project is compounded by manifold difficulties. First is the matter of urgency, the challenge to develop an ethics adequate to modern human power before that power destroys the planet. Second is the challenge of reconceptualizing the ground of ethics in a time suspicious of metaphysics. And third, in relation to this, is the task of restoring moral confidence in an age of massive moral confusion. On Jonas's view, these challenges need to and can be met by rejoining ontology and ethics, by reuniting being and value, by building an ethics up from the nature of things through a moral phenomenology of life.

As earlier described, one of Jonas's main methodological theses is that Darwin's theory of evolutionary continuity challenges the longstanding Western dualism of matter and mind. Because all of life's forms are related through a common history of natural processes, whatever capacities exist in life's highest forms must in some way be prefigured in less complex forms of life. Jonas writes, "The *continuity* of descent linking man with the animal world made it henceforth impossible to regard his mind, and mental phenomena in general, as the abrupt intrusion of an ontologically alien principle in the total stream of life."[26] In contrast to some ontologically alien principle, Jonas identifies the taproot of human freedom in life's shared mechanism of metabolism, as indicated in chapter one. All life metabolizes, and all life therefore manifests some measure of freedom and purposiveness and an inner horizon of transcendence. But Jonas does not equate continuity with sameness. Life is differentiated and the multiplicity of life's forms exhibit different degrees of freedom, different horizons of transcendence.

While life in general metabolizes, nature's surprise is that the "accident of terrestrial conditions brings about an entirely new possibility of being: systems of matter that are unities of a manifold . . . by virtue of themselves . . . ."[27] The organic as such, plant and animal life, is "actively self-integrating; form is not the result but the cause of the material collections in which it successively exists."[28] This thesis of life's self-integration warrants, for Jonas, the ontological conception of each distinctive form of life as an individual or subject. And yet the freedom that all life

26. Jonas, *Mortality and Morality*, 63.
27. Ibid., 65.
28. Ibid., 65–66.

expresses through metabolism is conditioned; in Jonas's terms, all freedom is "needful." Every organic unity is dependent on incorporating and using matter from the environment to sustain itself as an integrated unity. Every living form alters matter from the environment. Life qua life has the ability to alter matter, and this is life's rudimentary freedom, but life is also bound by a necessity to do this in order to live. Necessity adheres to freedom as a shadow to its figure.[29] This dialectic of need and freedom, of dependency and autonomy, provides the interpretive key to Jonas's account of organic differentiation.

The capacity to alter the material environment means that life as such is open and receptive to the world. Matter is a resource necessary to life that the organic in contrast to the inorganic has the capacity to manipulate. In critical opposition to Heidegger, this means for Jonas that even primitive forms of life "have world." "'World,'" writes Jonas, "is there from the very beginning: a horizon opened up by the transcendence of need, which breaks the isolation of inner identity to embrace a circumference of vital relationship."[30] This "breaking" and "embracing" means that organic phenomena essentially include some faint echo of subjective awareness. Every living form is "needfully free" in the openness of its relation to the other of its material environment.

What has thus far been described is attributable, according to Jonas, to the whole of the organic world. But with animal life this description becomes more graphically manifest. Central to the scalar complexity of life for Jonas is the notion of "mediacy." Mediacy is the underlying differentiating principle of life's forms. Plant life, relative to animal life, is fixed in its needfully-free relationship to the material environment. Plants transform inorganic compounds into organic ones through photosynthesis. What is needed is drawn directly from or is directly supplied by the environment. The metabolic horizon of freedom in which this process occurs is relatively immediate to plant life. The plant and the environment sustaining it are contiguous. But they cannot be conflated. For the form of the individual plant endures while its contents are in constant flux. The character of transcendence correlate to vegetative life, then, consists of the irreducibility of the individual plant to its constituent material. With animal life, however,

29. Ibid., 68.
30. Ibid., 69.

the gap between need and satisfaction spreads, the horizon of freedom expands spatially and temporally, and the character of transcendence becomes more pronounced. In contrast to plant life, animal life "is dependent upon the unguaranteed presence [spatial and temporal] of highly specific organic bodies [food] that are not always available [here and now]."[31] This gap is the distance that distinctively animal capacities can uniquely bridge. One suggestive way of putting this might be to say that absence is the principal ecological niche for animals. The capacities Jonas identifies as effected by and negotiating this niche are motility, perception, and feeling. And these capacities are themselves cause of an expanding spatio-temporal horizon of freedom. The development of motility, perception, and feeling is thus shaped within the context of an increasingly mediated relation between living animal subjects and their object-world.

But what more specifically does Jonas mean by saying that the horizon of freedom for animal life expands spatially and temporally? As already noted, for animals, in contrast to plants, the gap between need and satisfaction is opened. The satisfaction of animal need is not given in the same way as for plants. What is needed is at a distance, crossed by motility and perception, a crossing driven by feeling. Motility and perception correspond to the spatial expansion of animal freedom. The powers to move and perceive construct space as a wedge within the animal organism's world-relation, yielding one of the ways in which the horizon of animal freedom is expanded. The power to feel, on the other hand, corresponds to the temporal expansion of animal freedom. It discloses not spatial distance between organism and object, but the temporal distance between purpose and attainment.

Jonas writes, "In both cases [the expanded spatial and temporal horizons of animal freedom], distance is revealed as well as bridged: perception presents the object as 'not here but over there'; appetite presents the goal as 'not yet but to be expected.' Motility, steered by perception and driven by appetite, transforms *there* into *here* and *not yet* into *now*."[32] This is the essence of the predator-prey relationship. The animal needs either to capture or flee. The feeling of need motivates the response of either flight or pursuit. Jonas writes, "The span between start

31. Ibid., 72.
32. Ibid., 71–72.

and capture represented by this series of actions must be bridged by continual emotional intent."[33] Appetite motivates movement, and this is "the first sign of the difference between animal and plant: the placing of *distance* between drive and fulfillment, i.e., the possibility of reaching a distant goal."[34] Appetite, as desire or fear, maintains the goal-directed structure of motility and perception. The movement toward and away is motivated by either desire or fear, instantiating appetite as the basic form of animal self-concern.

In sum, the distance revealed and bridged by the animal capacities of motility, perception, and feeling does not exist for the plant. And through this distance emerges the subject-object split, a split that "is at the root of the entire phenomenon of animality and of its divergence from the vegetative form of life."[35] The capacities that simultaneously open this split and enable the crossing of the distance between the animal and its need are the source of the "greater and more perilous" freedom of animals.[36] This freedom is greater than the plant's because the capacities that leverage open the animal's world-relation mean that the animal is comparatively more independent; at the same time, it is more perilous insofar as the animal becomes by way of its independence much more vulnerable.

Given these comments, it is clear that by claiming through the thesis of evolutionary continuity that the organic as such harbors a capacity for transcendence Jonas is not implying organic sameness. The organic is differentiated, though all living things share metabolic processes in common. Metabolism is the common ground in which the seeds of biotic difference begin to flourish. There are crucial differences between plant and animal life, but the most profound of differences is that between humans and other animals. While through metabolism all life harbors transcendence and is free, this is germinal with respect to the freedom and transcendence of human life, as the following section will explore.

33. Ibid., 71.
34. Ibid.
35. Ibid.
36. Ibid.

Anthropos

In addition to the dialectic of necessity and freedom, Jonas also suggests "perception" and "action" as a related set of terms helpful to understanding the scale of organic differentiation. These terms, he claims, are aspects of knowledge and power, and the scale of organic difference reflects a movement from trace degrees of knowledge and power to their highest concrete embodiment in the human. Jonas writes:

> As for 'knowledge,' we see development in the breadth and clarity of experience and in increasing degrees of sensuous world-awareness, which lead via the animal realm to the most comprehensive and freest objectification of the totality of Being, found in man. As for 'power,' parallel to 'knowledge' and similarly reaching its pinnacle in man, we witness growth in the extent and manner of impact on the world—in other words, progressive degrees of freedom of action.[37]

Of all living forms, humans are the most profoundly free and powerful, and this freedom and power is a correlate of the unique ways that humans knowingly perceive themselves and act within the world they inhabit. In what follows, I will trace what Jonas takes to be several of the relatively distinct coordinates of a biologically informed philosophical anthropology.

While Jonas affirms the Darwinian thesis of evolutionary continuity and its challenge to dualism, viewing the theses of evolutionary biology as *necessary* to an understanding of the human, he does not assent to the *sufficiency* of these to a fully robust anthropology. Anthropology is more than biology. He claims that any explanation of the human that reduces the meaning of the human to evolutionary biological origins commits the genetic fallacy. The corrective to dualism cannot be a reductive monism. Each of these strategies produces a partial, distorted anthropology. Both miss the human as a psychophysical unity. Jonas argues that the capacities taken to characterize the human among other animals cannot fully be understood simply as survival mechanisms. It is a category mistake to confuse an explanation of the origins of these capacities for a full understanding of their meaning.

Thus, while Jonas is no friend of dualism, he does not believe the corrective to dualism is to go to the other extreme. Both dualism and

37. Ibid.

monism fail adequately to account for the similarities and differences between humans and our animal kin. In light of this position of Jonas's, he views his anthropological task as that of navigating the Scylla and Charybdis of dualism and monism. "In order to find the golden mean between [these] extremes," he writes, "it is time—and the task of a philosophical anthropology—to give thought to what is essentially beyond the animal in man without denying the features common to both."[38]

As with his differentiation between plant and animal life, Jonas's pursuit of his golden mean is directed by the concept of mediacy. What goes beyond the animal in the human needs to be identified as "a new stage of mediate relationship to the world that is already beginning to take form in animals and, in turn, is already based upon the mediate nature of all organic existence as such."[39] Through the concept of mediacy, Jonas articulates an anthropology that aims to account for significant moral and existential differences between humans and other animals while not losing sight of similarities. This aim flows from his affirmation that, in the words of one of his interpreters, between the human and other animals there are only "provincial" differences set within more pervasive commonalities, and that both these differences and commonalities are descriptively and normatively consequential.[40]

In keeping with the statement in the introduction to this chapter that one of the central themes of Jonas's moral anthropology is the connection between *being* and *doing*, an important place to begin to unravel his anthropology is with his interpretation of the human artifacts of tools, images, and graves.[41] Each of these artifacts signals distinctive human activity. They serve as heuristics to the human difference among other animals insofar as they prompt the questions of what conditions

---

38. Jonas, *Mortality and Morality*, 77.
39. Ibid.
40. See Donnelley, "Bioethical Troubles," 21–30.
41. It is instructive to acknowledge here that Jonas's selection of these artifacts is another instance in which Heidegger's philosophy seems to bear a strong influence on him. As I will eventually indicate, Jonas interprets tool-use as the origin of technology, image-making as the origin of metaphysics, and the ritualization of death through graves as the origin of ethics. Jonas's interpretation of these artifacts roughly parallels Heidegger's philosophical speculations on the themes of technology, language, and death. The crucial difference, as suggested in ch. 1, is that Jonas's treatment of these artifacts, and the capacities he takes their usage to imply, emphasizes their naturalistic underpinnings.

and capacities must be present within the kind of animal that produces them. Their existence in the world registers a certain mode of being, a particular kind of organismic mediation. If human being and doing are intimately connected, then what the human does is a clue to what the human is. If these artifacts are unique to the human animal, they can provide a window that looks onto the human difference within the larger province of animality. Jonas's argument is that these artifacts are indeed unique and thus an interpretation of them serves in the effort to delineate the human difference.

First, tools. For Jonas there is an important difference between the *implements* chimpanzees have been observed to use and human *tools*. Whereas a chimpanzee might use a stick to penetrate a termite mound in order to eat, the particular stick used by the chimp is for Jonas not fully a tool since it may not be used recurrently or held in reserve for future utility. A tool, for Jonas, is an instrument used for work that is itself worked on and made available for recurrent use. Summarizing what the tool suggests about humans, Jonas writes, "The *tool* tells us that a being, forced to deal with matter out of need, meets this need in an artificially mediated way that depends on invention . . . ."[42]

Jonas suggests that the difference between tools and implements reflects human eidetic capacities, the cognitive and imaginative capacity of the human to separate form from matter. A tool's form is first imagined as a means to some needed or desired end, then shaped out of matter in accordance with that form in order to attain some end. The difference between human tools and the implements used by other animals is a subtle and contestable one. Jonas acknowledges this by saying that the boundary between the two kinds of objects is fluid. The use of tools, in other words, does not mark a radical divide between humans and other animals. One of the significant points he wants to make here is the connection between human tools and the human capacity for making images.

Image-making reflects a capacity that "displays a total, rather than a gradual, divergence" from animals. Unlike tools, images for Jonas are in general biologically inconsequential. "Animal artifacts [such as termite mounds or spider webs]," writes Jonas, "have direct physical

---

42. Jonas, *Mortality and Morality*, 85.

application to the pursuit of vital ends . . . ."[43] But the construction of images does not directly or in all cases serve physical or biological necessities. Images reflect a level of mediacy between the human and the world that is peculiarly human, though it arises through the continuity of evolutionary processes. Making images entails separating form from the matter of the physical environment, and engaging the world in the abstract. The nature of an image as representation implies certain unique human capacities. For example, in contrast to mere perception of some object in the world, representational depiction entails intentional selection and stylization, which in turn imply an intensified mediation in at least two ways. First, the depicted image is intentionally abstracted or selected from the world, implying the cognitive capacity in the image-maker to recognize difference between the world and image, the symbolic capacity communicatively to play with or stylize this difference, and the physical capacities or controlled motor skills needed to make images. Second, and implied in the above, intentional abstraction belies a distance between the representing subject and the object represented.

An image is intentionally depicted as both similar to and different from the object in the world it represents. Selective and stylized depiction represents likeness, not identity. And likeness turns on "incompleteness," otherwise an image would not be an image but a reduplication of some actual object.[44] So an image as image is incomplete with respect to the object in the world, and the selective omissions arranging this incompleteness are contingent on a high degree of epistemic mediacy between the image-maker and the world. The particular object in the world is represented in generalized form through the omission of certain details of the particular. In addition to omitting details, an image may intentionally alter the features of the object being represented. Intentional alteration, if it does not stray too much, can increase the symbolic work of the image.

The image is recognized *as* an image. It is grasped as distinct from reality and thereby becomes a new object under the control of human intention. The image comes under the control of the subject through the exercise of imagination, and signifies at least two features of the

---

43. Ibid., 79.
44. Jonas, *Phenomenon of Life*, 159.

human's expanded horizon of freedom: "The freedom gained in this manner—to ponder things in the imagination—is one based upon both distance and mastery."[45] Distance is embedded in the very nature of an image to the extent that an image externalizes some object abstracted from the world and allows it to be reflectively held in the imagination. Mastery includes reciprocally operative physical and mental dimensions. Physically, Jonas notes, the capacity to make an image entails a command of bodily motor skills not evident among other animals. Jonas refers to the relation between the physical and mental dimensions of mastery by commenting that, "the eidetic control of motility, i.e., muscular action governed not by set stimulus-and-response patterns [is] freely chosen, inwardly imagined, and [results in a] purposely projected form."[46] The human image-maker, *homo pictor*, integrates the external bodily mastery of *homo faber* and the inward imaginative mastery of *homo sapiens*. In sum, "The *image* tells us that a being, using a tool on matter for an immaterial purpose, depicts the contents of his visual perception, varies them and transforms them, thereby creating a new world of depicted objects beyond the material world that is there to satisfy his needs."[47]

And so for Jonas one of the distinctly characteristic features of human life in contrast to other animals is that humans construct and imaginatively inhabit a second, symbolic world that signals a more-than-biological dimension to human existence. With the image-making capacity, one aspect of the human world-relation can be characterized as an internalization of the external. This inner duplication of the outer is largely under human control, in contrast to the outer world's largely uncontrollable nature. But the inner world's mediation of our relation to the external world ultimately makes possible a new degree of leverage over the outside. For through the image, the world is not only reproduced. Imaging can alter the world through its effect on human consciousness and the bearing of consciousness on human conduct. As I will later explore, there are profound ethical and theological implications in this idea. For, as I interpret Jonas, image-making is not only a distinct human capacity but also a burden of moral responsibility.

45. Jonas, *Mortality and Morality*, 81.
46. Ibid., 82.
47. Ibid., 85.

The grave, another artifact Jonas explores in his quest for the anthropological golden mean, becomes for him an especially significant mark of the uniquely human level of mediacy. The analysis of tool and image resulted in an appreciation for the human's heightened mediation from the world. The grave reflects human self-mediation and signifies another level of mediacy. Unlike other animals, the human knows he must die, and this consciousness presses him to reflect existentially on his identity, purpose, and place within his world-situation. Jonas claims that as early tools are the primitive forms of the technology, as image-making is at the origin of art and metaphysics, the phenomenon of the grave is at the source of ethics. This is so for the fact that the existential questions produced by consciousness of death result in the construction of a normative self-image. Awareness of a limited future forces reflection on the human's present situation, which reflection produces an externalized self-image. This externalized self, the "I," becomes another of the objects within the symbolic world the human inhabits. Whether the human affirms or rejects the self-image, whether or not his inner life and behavior accord with the image, the image operates as a normative standard.

So like other animals, the human "has world," but unlike other animals the human also "has himself." The empirical fact that only humans create graves testifies to the idea that it "is not only man's relationship to the world which is [mediated] but also his relationship to himself. He arrives at his own being only via the detour of ideas about it."[48] Speculation feeds here in the gap between the self itself and its image, nourishing a new entity among life's phenomena. Organic mediacy is most pronounced here in the unique human capacity to imagine life against the pressure of mortality, in short, to have not only a "world" but also an "I." Jonas writes, "With [the grave] this attainment of distance and the bridging of it by means of never-ending reflection, the principle of mediacy, with which life began and whose growth can be traced through all of organic evolution, reaches its pinnacle."[49]

The height of the arc of mediacy, curving from metabolic processes toward animal motility, perception, and emotion is occupied by humans whose distinctive capacities for making tools, images, and graves

---

48. Ibid., 84.
49. Ibid.

signify our self-mediation and our unique distance from and control over the world. The paradox that this insight yields, when considered against Jonas's interpretation of technos, is that the world that has generated this creature becomes by that very generative process vulnerable to demise. This vulnerability stems from Jonas's position that our world is fundamentally a technological one, threatened and defined more than any other thing by the hegemony of the tool. He writes:

> Our culture today places the greatest emphasis on what was foreshadowed in the tool: technology and the natural sciences that serve it. Tools, which—of the three—best served the purpose of biology and its dynamics of selection, first appeared in response to the constraints of nature. Continually surpassing themselves with their undreamed-of successes in recent times, they now completely dominate our entire external existence, overshadowing everything else that distinguishes us 'from all beings that we know' (Goethe).[50]

Protecting the planet from demise at the hands of *homo faber*, not to mention the human from his own hands, depends for Jonas most essentially on the normative guidance to be offered by a moral anthropology that includes more of what distinguishes humans from other animals. Tools, as Jonas claims, mark merely a fluid boundary within animality. Images and graves, on the other hand, denote much more significant differences, the biologically novel mediacy of the human self-relation and a radically heightened mediacy between the human organism and the world. These differences need to be balanced against the present dominance of the tool, for they are ultimately, for Jonas, the sources for the moral vision and practice necessary to the normative direction of human power.

Against the thinking of many environmental ethicists, then, Jonas argues that while human continuity with nature should not be neglected, what the planet needs most are humans that are more fully human and not less. Paradoxically, this ethical charge entails a return to what human animals share with all other forms of life—metabolism. This return, though, will be one that more intentionally excavates its ethical implications. In what follows, I will examine the connective

---

50. Ibid., 86.

tissue whereby Jonas unites metabolism and organic identity with his argument for the ontological ground of value.

As interpreted above, graves provide a significant clue for Jonas to the unique degree of mediacy of human animals. They signal the fact that humans live with a consciousness of mortality. Under the pressure of this consciousness arises the uniquely organic phenomenon of self-mediation, for awareness of death's proximity prompts existential reflection. This reflection in turn yields the production of a self-image that can serve as a normative existential standard. Relating to this image, we measure ourselves as we are against what we would like to be. While normative self-imaging is uniquely human, mortality of course is not. Death is inevitably linked to life. Mortality is the condition of organic nature—it is life's constant possibility. The animate is a temporarily organized wakefulness of matter that always in time returns to matter's slumber. As Jonas puts this, "Life . . . carries death within itself."[51] Another way of putting this that brings it together with the metabolic dialectic of freedom and necessity is to say that organic nature is simultaneously in opposition to and solidarity with inorganic nature. It always depends upon a solidarity with and transcendent opposition to inorganic matter. However this is expressed, if mortality is life's constant possibility, then it is integral to the phenomenon of life, whether consciously experienced as such or not.[52]

This fairly obvious point about the inevitably of death for the living begins to unveil the meaning of Jonas's more opaque idea, alluded to in the introduction of this chapter, that for the organic, being is doing. Organic being depends not only on the continuous task of doing, but also on the environment that serves as the field of the organism's doing and supplies the materials of its activity. While its being is its doing, the organic is dependent as well on the "hospitality" of its environing world.[53]

---

51. Jonas, *Mortality and Morality*, 90. First presented to the Royal Palace Foundation in Amsterdam in March 1991, then delivered at a conference honoring Jonas at Hebrew University, Jerusalem, January 1992.

52. This, again, illustrates one of the ways in which Jonas reinterprets the ideas of Heidegger.

53. My usage of "hospitality" alludes to Leon Kass's treatment of the concept in *The Hungry Soul*.

This language is simply another way of talking about metabolism, but a way that intentionally defamiliarizes it in order better to draw out its ethical significance. Metabolism is the doing of all and only organic beings. It is the doing of which the cessation is death. In the whole of the natural world, only living beings endure by way of such a precarious and paradoxical process. The inorganic endures ineluctably—it simply is. The identity of particular inorganic composite things consists of the relative stasis of their ingredient elements. The identity of organic beings, cannot, for Jonas, be understood this way. The matter of which an organic entity is made is in constant flux. For example, while a particular hare seen on one day can be identified as the same hare on another, given sufficient time the material of which it is made will be completely different. "Repeated inspections [of a living organism]," Jonas writes, "find it to consist less and less of the initial components, more and more of new ones of the same kind that have taken their place, until the two compared states have perhaps no components in common anymore."[54] So the identity of organic entities is not equal to their constituent material, and yet without the continuous metabolic incorporation, use, and excretion of matter, the organic being ceases to be.

The identities of living things are not the same as their ingredient materials, but they are dependent on the ongoing activity of securing and processing these materials. Different levels within the organic—plant, animal, and human animal—are characterized by their different modes of activity. The plant is "given" the matter it needs, and the activity of processing it is strictly internal. With animals, activity takes on an externalized character. For the vital material to be metabolized is not given to the animal, but needs to be secured—animal activity is separate from its goal or purpose, the attainment of vital need. The result of the animal's more mediated relation to the world is a picture of the animal as an "isolated individual pitted against" a world that is "at once inviting and threatening."[55] In a more pronounced way than for the plant, survival for the animal is an issue of conduct in the world.

But at whatever level, the relation between metabolism and organic identity is a paradox. As Jonas puts this:

---

54. Jonas, *Mortality and Morality*, 89.
55. Jonas, *Phenomenon of Life*, 105.

> We are faced with the ontological fact of an identity totally different from inert physical identity, and yet grounded in transactions among items of that simple identity ... the living body is a composite of matter, and at any one time its reality totally coincides with its contemporary stuff ... [and at the same time] it is not identical with this or any such simultaneous total, as this is forever vanishing downstream in the flow of exchange ....[56]

The paradox here concerns the dialectic between independence and dependence, freedom and necessity. If the identity of the organic individual is not the same as its ingredient materials, then it is in some measure free from matter. But if at the same time the organism depends on these materials to sustain itself, then it also exists in a needful relation to matter. Thus, whether plant or animal, organic identity is constituted by the simultaneity of freedom from and dependence on matter. Organic existence is fundamentally precarious. Proper interpretation of the organic requires taking in the whole of this relationship, privileging neither need nor freedom.

Jonas's views here on mortality and the paradoxical linkage between the needful freedom of metabolism and organic identity provide the rudiments of his moral anthropology. In place of the either/or of dualism and monism, he describes a constitutive dialectic at the core of life. The anthropological significance of this dialectic and the way in which it seeds his larger normative vision are freighted within the paradox of identity. For through his differentiation of life, he traces the biological ascent of freedom, an ascent that reaches its summit in the human. This freedom is specifically human and yet in continuity with the rest of life, differing in degree from the freedom of other organisms. It is through both sides of the biological dialectic—necessity and freedom—that the identity of human animals comes to be especially characterized in terms of moral responsibility. It is this moral nature of human freedom, rather than freedom as such, that designates the crucial qualitative difference between humans and other animals.

Jonas's organic equation yields the position that organic identity is not the same as the product of its parts. Nor is it merely a function of the metabolic calculus. Though composites of parts, organisms are more than their sum. Though fueled by the metabolic calculus, organisms are more than its function. As more than the effect or product of metabo-

---

56. Jonas, *Mortality and Morality*, 89.

lism, effective causality offers only a partial, even if a crucial, explanation of the organic. Nor is material causality a sufficient explanatory path. Understanding organisms cannot be reduced either to descriptions of their ingredient matter or to the mechanics of metabolism. For if organismic being is organismic doing, rather than merely the product of doing—or if organismic being depends on metabolism and yet at the same time is more than a function of it—then any particular organism is in some way its own formal cause. And, further, organismic self-concern with continuing to be is their final cause. Organismic identity is not reducible to either the matter or effect of organismic doing. The matter and mechanics of metabolism sustain the organism, but organismic form and telos themselves sustain this sustaining process.

These ideas are crucial to understanding Jonas. They reflect a reinterpretation of a physicalist or materialist explanatory strategy on its own terms, allowing Jonas to gain leverage on his critique of the inadequacy of the modern scientific view of a natural world indifferent to itself and the life forms it generates. Organisms are self-integrating beings whose telos is to continue to be. In other words, the nature of the organic is to be auto-telic, self-constituting, self-regulating, and metabolism describes only effective and material aspect of this causal process—organic identity is related but not reducible to the parts, structure, or processes that inform it.

What holds for the organic as such holds emphatically so for humans; for the human animal stands, at least for the present, at the apex of biotic evolution. Jonas's phenomenology of life yields the insight that organic form and telos play causal roles in organismic being, and that this increases in concert with increasing degrees of mediacy. But with humans, according to Jonas, form is not merely causally self-generative and the human telos is not merely that of continuing to be. Instead, to be human is to be responsible.

As I interpret Jonas on this score, the human form is responsibility, and the human telos is to be responsible. I take this to be the case for three interrelated reasons. That the human form is responsibility and the human telos is to be responsible are the consequents of, first, the biologically extraordinary powers of human efficacy, and, second, the human capacity for self-awareness. The distinctiveness of human efficacy is related to our intensely mediated relation to the world, evidenced in the tool. Self-awareness, a correlate of human self-mediation, is evidenced in the image and grave. Together, efficacy and awareness provide the antecedent conditions for the meaning of responsibility.

Humans are not only effective beings, but also self-consciously effective. Humans are aware of their impact on the world, of their causal accountability. With these two points, however, the moral specification of responsibility has not been fully accounted for. The moral quality of human responsibility depends on linking human efficacy and awareness to value. That humans can impact the world and are aware of their capacities to do so means that conscious causal responsibility can be attributed to humans, but not necessarily moral responsibility.

The third reason in support of the idea that the human form is responsibility and the human telos is being responsible, and for characterizing this responsibility as moral, is derived through Jonas's ethical excavation of metabolism. Insofar as organisms metabolize, and metabolism is primitively the affirmation of being against nonbeing, an expression of life's choice for itself, then organisms are purposive even when those purposes are not consciously held in view. And given this innate purposiveness, every individual organism has value for itself.[57] For, according to Jonas, the attribution of value is simply a logical entailment of organic purposiveness—the pursuit of any end or purpose denotes an infrastructure of value. Purposiveness is present wherever there is an organism, and where there are purposes there is value.

The moral character of human responsibility comes into view at this point. Humans become morally responsible when conscious efficacy impinges on a world of beings with value. Thus the human form is moral as well as self-generative, and this is a consequent of human efficacy and human awareness of value. The human telos is to be morally responsible, and this is a consequent of the idea that while the human

---

57. As has been shown, the difference between the inorganic and the organic is constituted by the emergence of a dialectic between affirmation and negation. Organic being only comes to be and only continues to be against the possibility and threat of not-being. The "doing" without which organic "being" cannot continue to be reflects life's fundamental affirmation of itself against its negation, its concern to be rather than to not-be. Life is essentially a hazardous gamble. Jonas deduces a moral fact from life's gambled choice of itself, from life's assertion of a "yes" against the "no" of not-being—that life fundamentally is valuable. These ideas illustrate the extent to which Jonas has reworked Heidegger's ideas. The "being-toward-death" that Heidegger marks as peculiar to Dasein is on Jonas's interpretation a feature of all life, and the genesis for value. Jonas's view is not that all life lives "consciously" toward death, for that is unique to human life, as signaled by the phenomenon of "grave-making."

telos includes the essential organic will to continue to be, it is more than this. As a result of humanity's radical world-altering capacities, the horizon of human doing impacts not only human beings, but also many other beings deeply imbued with value, and in our time, even the whole of being.

Conclusion

As earlier interpreted, Jonas's work is driven by the view that the unique moral demands of the present world require a reunion of ethics and ontology, a theory of the good rooted in the soil of bring or the nature of things. I began to trace Jonas's treatment of the eclipse of being and value in chapter one, with an interpretation of his views on Gnosticism and Heidegger's philosophy. This chapter's section on *technos*, the ethos of the contemporary world, completed his account of this eclipse.

Recall that *technos* is the result of a confluence of conceptual revolutions and their practical moral consequences. The image of the world and humanity pictured by this confluence is one in which human moral purposes are adrift in an indifferent universe. The unfolding historical effect of this image, for Jonas, is the moral ambiguity of human power. For according to such a picture, nature does not provide any norms for human conduct. The only source for norms becomes human power itself. Value is reduced to human invention. And thus the historical narratives underlying *technos* coalesce to show how the severance of being and value is both cause and effect of increasing human alienation from the natural world, the dominance of the technological enterprise, and the rise of human power as its own end and as the dominant source of moral worth.

Though technos, organos, and anthropos designate different dimensions within Jonas's view of the moral world, they are united by a common theme. This theme is the constitutive reciprocity of *being* and *doing*: the being of all living forms is structured actively as doing. While efficacy varies among life's forms, it is essentially purposive. Doing is the way that organic being sustains, and in sustaining, affirms itself against nonbeing. This native biotic purposiveness serves Jonas as a necessary condition for the reunion of being and value. This is so since for Jonas any purposive movement toward some end instantiates that end's subjective value in relation to the organism pursuing it. Value is present in the purposive activity of individual organisms.

And thus the counter-image of the world and humanity that Jonas develops, with the aid of the modern natural sciences, presents a picture of a world of living things that is purposive by nature and that harbors intrinsic value. Life forms evolve in continuity with one another, but display increasing degrees of mediacy. The form and telos of human moral responsibility marks a crucial difference between humans and other animals—but not a total cleavage. For responsibility is rooted in the theme of mediacy that runs through the organic, the theme of increasing difference from and power to alter the world.

This theme merely reaches its crescendo in humans, though it is a piercing crescendo. The mediated relation between the human and the world is so dramatically intensified that it comes eventually to structure the human self-relation. The "I" becomes an image within the symbolic world in which the human lives in simultaneity with the concrete physical environment. The human undertakes the project of existence not only through a highly mediated world-relation, but also in relation to a self-image projected under the pressures of a unique consciousness of mortality and an awareness of the vulnerable purposiveness and value of other forms of life.

Unlike any other, the human organism is self-mediated, and this mediated self-relation is at the origin of ethics for Jonas. As no organism can fully transcend need, no matter what its degree of freedom, the human can never escape its own image. Jonas writes, "Henceforth, like it or not, man—each one of us—must live the idea or 'image' of man, an image that is constantly being modified. It never leaves him, however much he sometimes yearns for the animal felicity he has lost."[58] As chapter five will show, Jonas holds that "the image of man" is not only inescapable, but also that the preservation of it is the greatest task of human moral responsibility.

By tracing the ascent of freedom through the organic, by specifying human freedom as moral through the artifacts of tool, image, and grave and in relation to a world of living things dense with value, Jonas's moral anthropology is characterized well by the root metaphor of responsibility. Within the phenomenon of life, humans are uniquely

---

58. Jonas, *Mortality and Morality*, 84.

able to behold the purposive form of things and thus are uniquely responsible for seeing and feeling the basis of value and the imperative to preserve it.

However, much remains to be done in order fully to grasp Jonas's understanding of the gravity and burden of human responsibility. For according to Jonas, the human is morally responsible not only for the being of individual others, but also for *all* being and especially for the future of the idea of responsible humanity. Understanding this depends on following Jonas's argument from the purposiveness and value of particular organisms, just explicated, to his argument in support of the value of purposiveness as such—in other words, from the ontic grounding of value in beings to the ontological grounding of good. This will be taken up more fully in ch. 5.

In closing this chapter, I would like to underscore that Jonas's accent on freedom is crucial to his deepest anthropological metaphor—responsibility. This emphasis here of Jonas's is significantly different from Gustafson's. Gustafson's deepest moral anthropological metaphor, participation, assumes a different account of the character of life and the broader nature of moral reality. There is much at stake ethically in these alternative anthropologies and their attendant moral visions.

With Jonas, the constituent category of moral reality is the living individual, while with Gustafson the field of moral reality is a more intricately textured whole that contains theological dimensions as well. In contrast to Jonas, Gustafson attends not so much to living individuals as to the patterns and processes underlying and linking individuals one to another and to the divine. My point in noting these differences is to suggest that they present two equally important insights into the situation of human life within a complex world—that humans, as individuals and collectively, uniquely stand out in the natural world in addition to being participants embedded within it. My aim is to show that, for descriptive and moral reasons, it is important to attempt to hold these insights together in dialectical embrace. It is now time to turn to the fundamental ideas in Gustafson's vision of moral reality, in order to continue to interpret what he can offer, in complement to Jonas, to an ecotheological ethics of responsible participation.

4

God, World, Human Being

Introduction

THE STARKEST CONTRAST BETWEEN JONAS'S AND GUSTAFSON'S FUNdamental dimensions is theological. Thus far, the role of theology in Jonas's work has not been engaged in detail, though it was mentioned earlier that he views theology as a "luxury of reason." He argues that the descriptive and prescriptive claims he makes do not require theological backing, but neither are they incompatible with certain theological ideas. While the meaning of this will be interpreted more fully in the next chapter, it is clearly different from Gustafson's understanding of theology.

Gustafson argues that insofar as every ethical theory presumes some account of the way things really are, every theory depends upon a theological vision or a theological surrogate. This of course is not always made explicit, but according to Gustafson every moral position includes foundational premises that either are theological or, in terms of their ultimacy, are functionally equivalent.

In articulating a theological framework for his ethics, then, Gustafson does not consider himself to be doing what a theologian must do, but instead to be making explicit what every ethicist always does. As theologian Gordon Kaufman puts this, "For Gustafson God is not an extra and dispensable reality of interest only to 'faith'; rather *God* is our name for the ultimate reality with which humans must come to terms in life . . . ."[1] God, for Gustafson, is the name for the power ultimately ordering the world and bearing down upon and sustaining

---

1. Kaufman, "How is God to Be Understood in a Theocentric Ethics?" 16.

it, the power that structures all reality and is therefore indispensable to ethical theory construction.

While Gustafson's project is emphatically theological, he does not move deductively from theological positions to other kinds of claims. Though fundamental in his work, theological principles do not have methodological priority. Instead, as the chapter on his hermeneutical strategy showed, Gustafson grants methodological priority to experience. Common to human experience, as Gustafson understands it, are several universal senses. These need not but can be construed as religious.

Gustafson's correlate understanding of theology is that it is a second-order enterprise of critical reflection on religious experience. At the same time that he argues this, however, he also affirms that all conceptualizations and expressions of the significance of experience are always tradition-mediated. Thus any theological construal of the significance of experience may already pre-reflect the emphases of particular theological traditions. This appears to present an epistemological circle.

As discussed in ch. 2, Gustafson works to avoid this circularity by arguing that in addition to referencing tradition-internal criteria of meaningfulness, experience also should be articulated with reference to more public criteria of intelligibility and persuasiveness. The significance of any experience, any phenomenon, any text or event can be and often is articulated from multiple perspectives. All phenomena, in Gustafson's view, are discursive "intersections."[2] There are always multiple ways of seeing, describing, explaining, and evaluating things. In other words, there is "traffic" at every "intersection."

This provokes the question of what criteria should direct the traffic. To what criteria of justification should one appeal in rejecting, assimilating, or modifying the perspective of some one of the various possible ways of construing things? The answer to this question depends, to no small degree, on the primary discourse with which a thinker works, for different discourses abide different canons of meaningfulness. But how different are they? Are they incommensurable? Are they overlapping, can they be integrated? In arguing that theological construals of experience *should* be intelligible and persuasive outside particular theological

---

2. See Gustafson, *An Examined Faith*, 5.

communities, Gustafson is also suggesting that they *can* be. While theological and nontheological discourses are different, they *can* speak to one another on Gustafson's assumption that there is a base standard of reference to which they each can appeal—common human experience. Further, according to Gustafson's logic, theological and nontheological discourses *should* speak to one another if the density of any experience, event, or phenomenon is to be adequately accounted for.

These methodological points, examined in more detail earlier, do not come from nowhere. Gustafson's prescribed hermeneutical to-and-fro between experience and the articulation of its significance, and the normative consequences that follow from it, reflect his fundamental commitments. In keeping with my own methodological claims, the hermeneutical, fundamental, and normative dimensions of Gustafson's project interpenetrate.

This chapter will focus on Gustafson's fundamental ideas of *God*, *world*, and *human being* and consider the way in which they point back to his hermeneutics and forward to his moral theory. The aim, first, is to interpret the reasons behind Gustafson's critical revisions of the Reformed Christian theological tradition with which he identifies. I will then trace the impact of these revisions on his fundamental ideas. As insightful as Gustafson's theological revisions may be, in my conclusion I will critically treat the question raised by many of his interpreters, the question of whether his theological construal of things really adds anything to a more strictly naturalistic account of the way things really are. But in order to advance to an assessment of this, it is crucial first to consider the good reasons for Gustafson's theological revisions.

A Revised Reformed Theology

Gustafson's understanding of the Reformed tradition is genealogically capacious, extending backward to Augustine and forward to John Calvin, Jonathan Edwards, and Friedrich Schleirmacher. Though there are of course significant differences among these theologians, Gustafson identifies three basic emphases they share that allow him to speak of their work as an identifiable tradition. First, they commonly emphasize God's otherness and sovereignty; second, they focus on the attitudes of reverence and piety stemming from this sense of otherness; and third, they share a basic conviction that right human action is action correlated to the purposes of God.

It is not necessary here to consider fully Gustafson's interpretation of the major interlocutors within this tradition. The aim at this point is simply to sketch the theological framework through which he construes his fundamental ideas. Basic to this framework, as the following will show, are his critical challenges to the tendency within the Reformed tradition to anthropomorphize God, the resulting anthropocentric axiology that follows from this, and the tradition's problematic account of moral motivation.

Gustafson's critical revisions to the Reformed tradition begin, first, with a challenge to its anthropomorphic understanding of God—its tendency to view God in personalistic or agential terms. In addition to tradition-external demands for this challenge, Gustafson deems that such a challenge is required if the tradition is to be internally consistent. On this score, Gustafson argues that a doctrine of God that specifies the divine nature according to human capacities, however perfected those capacities may be expressed, is not compatible with the tradition's emphasis on divine sovereignty and otherness.

Divine sovereignty often is described as being determinative of both the general and particular events in nature and history and of limiting human control over these events. Tethered to divine determination is an emphasis on divine foreknowledge. The image of God rooted in human attributes derives from the combination of these two emphases. God is imagined as the supreme agent directing or governing the world with intelligence, will, and purpose. The anthropomorphism of this agential image is to some degree the inevitable result of the analogical reasoning intrinsic within much of Christian theological discourse. Analogical reasoning moves from the assessment of human limitation to divine unlimitation through the presumption of God's sovereignty and perfection, and in this way, "Criteria from *human* life are the basis for attributes of Divinity."[3]

The anthropomorphic indicative of God shapes the Reformed theological tendency, indeed, the tendency in Gustafson's judgment of most Christian theological discourse, to characterize the divine imperative or divine purpose in anthropocentric terms. God is not only

3. Gustafson, *Ethics from a Theocentric Perspective*, 1:179.

too often conceived in human agential terms, according to Gustafson, but the human is too often conceived as the center of value in God's creation. Characterizing this is "the claim that the divine determination of events occurs for the sake of human well-being, or on the basis of human deserts, or deserving."[4] The Reformed tradition, along with Christianity more generally, tends to assume that humans are the principal objects of divine beneficence.

Underlying this particular assumption is the pervasive philosophical and theological view that humans stand near the highest rung on the ladder of being. On this understanding, all that occurs through nature has meaning, significance, and value principally in relation to the human species. Theologically, everything that happens in nature and history, positively and negatively, has meaning insofar as it reflects divine justice and beneficence toward humans. All events, those initiated by humans and those beyond human control, reflect the justice and goodness of divine judgment on humans. Even tragedy and the experiences of human suffering are ultimately intended for the human good according to the divine will.

As with an anthropomorphic idea of God, this anthropocentric rendering of divine purposes is problematic for Gustafson for theological, scientific, and experiential reasons. Gustafson's critique of theological anthropocentrism is reflected in his ironic summary of it:

> Thus everything that occurs either as a result of processes over which we have no human control or as a result of the exercise of human agency must be meaningful in relation to God's justice and goodness for our species if not explained by it—our species, so late in developing, living on this small planet within our solar system.

In spite of this critical tone, Gustafson affirms that God is for humans, but in a qualified way. While God may be for humans, God's endorsement need not nor always does accord with human metrics of goodness. The anthropocentric claim that God's purposes always accord with a human criterion of good is problematic with respect to common human experience as well as in the light of greater scientific understanding of the natural world—in which it is clear that humans are biotic latecomers occupying a very small niche in the greater cosmos.

---

4. Ibid., 180.

Gustafson also critiques the tradition's interpretations of the relation between moral motivation and the doctrine of eternal life. Eternal life is a seemingly necessary logical correlate of the attempt within the tradition to hold together divine justice and beneficence. Eternal reward is promised to those who are virtuous yet experience misery in life; eternal punishment is promised for those who experience good fortune beyond moral desert. Though the doctrine of eternity articulates an account of reward and punishment, it is not supposed to serve as the motivation for the good and right life before God. Gustafson's principal issue with this doctrine is not so much that there is no scientific evidence to support it, nor that it cannot be experientially confirmed, but that it tends in practice to become the motivating force of moral action and religious life. According to Gustafson, on the tradition's own view that the primary purpose of humans is to glorify God rather than to attain reward, the doctrine of eternal life needs to be rejected.[5]

In light of these critiques of anthropomorphism, anthropocentrism, and moral motivation, Gustafson, like Jonas, uses his tradition against itself in order to advance a reconstructive position. But it should be noted at this point that part of the fallout of his critical reconstruction is that it generates a significant problem for the tasks of moral reasoning. This problem is related to the radical epistemological gulf the tradition posits between the absolute otherness of God and limited human rational capacities. If right action is action in accord with divine purposes, and yet there is a gulf between the divine and the human, then to what degree can one hope ever to gain epistemological access to these purposes, let alone to have the capacities to act in accordance with them?

---

5. Gustafson acknowledges, "One point at which I have turned from a central aspect of the tradition is the displacement of the salvation of persons as the principal point of reference for religious piety and for the ordering of theological principles" (*Ethics from a Theocentric Perspective*, 1:112). Gustafson understands that rejection of such a central aspect of Christian theology cannot be undertaken lightly, that for some readers it might lead to the judgment that his position is not recognizably Christian. And yet he significantly challenges the underlying logic of the doctrines of salvation and eternal life, the constriction of the moral nature of the divine by human definitions of "moral." Above all, echoing Jonathan Edwards' view that the human moral and religious purpose is to glorify God, Gustafson's point is that "God does not exist simply for the service of man; man exists for the service of God" (*Ethics from a Theocentric Perspective*, 1:342).

Gustafson acknowledges this problem in his interpretation of human sin. Sin is an unmistakably central theme in Reformed theology. Though central within the tradition, Gustafson also claims that a sense of human fault or sin is culturally universal, as explained in chapter two. Because this sense is so pervasive, so also is the desire for forgiveness or relief from it.

For Gustafson, this sense of sin and the desire for relief fund the utilitarian-instrumentalist trends in many religions. Various strategies, techniques, and therapies for guilt relief often play a central role in religious beliefs and practices and are often spun instrumentally as the "profit" of religion. Since Gustafson wants to affirm human fault as a universal human sense and also to resist the instrumentalization of religion, the question he must face "is whether one can take the human fault with deep seriousness, establish some sense of the possibilities of human alteration, and claim some benefits of the divine benevolence, without becoming trapped in utilitarian Christianity's preoccupation with human guilt."[6] This question, essentially a question about the dynamic between human freedom and divine sovereignty, will be faced further along in the interpretation of Gustafson's moral anthropology. Having accounted in these previous paragraphs for his general affirmations and critiques of the Reformed tradition, it is now appropriate to turn to the ways his revisionary theological framework shape his fundamental ideas.

God

In his articulation of the idea of God, Gustafson clearly puts to work his hermeneutical criteria of scientific, experiential, and theological congruity. As earlier discussed, these criteria state, first, that the significant content of theology cannot be radically incongruous with scientific knowledge and, second, that theological claims should be congruent with human experience broadly construed. Together, these positions mean that particular Christian themes, symbols, or emphases may have to be abandoned or at least modified in the light of human experience and scientific knowledge. As Gustafson understands them, both theology and the sciences are developmental traditions, dynamically responsive to new knowledge and historical conditions.

6. Ibid., 185.

Gustafson attempts to honor his commitment to congruence through a strong theocentric reading of the classic Christian symbols of God, filtered through his theory of the natural senses. In sum, common human experience suggests that humans are radically dependent on powers beyond their control, that the ordering of these powers sustains the possibilities and determines the limits of life. Common human experience also testifies to the universal sense of transgression, but additionally points to the universal senses of possibility and direction. Gustafson affirms the senses of dependence, gratitude, obligation, remorse, possibility and direction as universal and common, but specifies them in relation to his own theological tradition through a reinterpretation of the classic Christian symbols of God as Creator, Sustainer, Judge, and Redeemer.

The central move in his reinterpretation consists of a switch from an anthropocentric to a theocentric mode of theological discourse. In keeping with the Reformed theological emphasis on divine sovereignty and otherness, but in his view, against the grain of common piety, Gustafson takes God to be the center of value rather than humanity. There is no necessary reason to assume that God's beneficence is always calibrated to what humans deem to be good when the claim that God is the center of value is combined with an emphasis on divine sovereignty. The theological defense of theocentrism against anthropocentrism is, according to Gustafson, further supported by natural scientific insight into the infinitesimally small place humans occupy in the space and history of the cosmos.[7] The central claim of Gustafson's theocentrism is that the world-ordering sovereign power on which humans depend is not concerned exclusively with the human good as humans generally understand it. From this theocentric rather than anthropocentric

---

7. For Richard McCormick, this description of the place of humans in the cosmos does not necessitate the dismissal of anthropocentrism. According to McCormick, "The fact (?) that the universe antedated human life by millions of years and will be around after life as we know it has disappeared can be read as God's lavish way of presenting human life precisely as the crown of the universe." See McCormick, "Gustafson's God" 57. In other words, the move Gustafson makes from a description of the place of humans in the history of the cosmos does not require, according to McCormick, an axiological devaluation of humans in the ordering of things. As I will explore in the next chapter, Jonas offers a different interpretation from Gustafson, indicating that the evolutionary lateness of human life is in fact at the source of humanity's greater value within and responsibility toward nature.

starting point, Gustafson moves into critiques of traditional Christian theological symbols of God.

God as Creator draws its symbolic strength for Gustafson from the sense and experience of dependence in a world not of human making. Along with the sense of dependence that the symbol Creator expresses are the senses of interdependence and interrelation with natural and social patterns and processes and the senses of possibility and direction that arise through an appreciation for human capacities. For Gustafson, the Genesis narratives "express human consent to the powers that have brought life into being; that order the range of 'objects' and experiences; that create the conditions of possibilities of human development biologically (genealogies), historically, socially, culturally."[8] The symbol of God as Creator, then, describes the common experience that the power or powers that created the world and the interdependent conditions of life transcend the human and can generate human consent to and awe before the power(s) that bring(s) life into being.

Next, Gustafson critiques the traditional concept of God as Sustainer for relying too heavily on the assumption of an immutable divine order. The immutability of divine order is affirmed in accordance with a classical view of the nature of God as unchanging and eternal. The idea of perfection is here aligned with changelessness and contributes to a scientifically and morally problematic account of the worlds of nature and history as static. Morally, the emphasis on stasis can tend to sanction "a conformity of institutions, historical events, and cultural achievements with this changeless, timeless divine order."[9] Gustafson's scientifically informed critique of stasis will be engaged more closely in the following section on his understanding of the world. But it is important to note here that in place of "order," Gustafson deliberately chooses to speak of "ordering." Viewing the Sustainer as "ordering" power better reflects for Gustafson the dynamism of natural processes described by the modern sciences. The symbol of God as Sustainer also derives from the common sense of gratitude for the way the world seems to conduce to the sustenance of life. The Sustainer symbol brings to consciousness that, "We continue to rely upon processes and relationships that are

---

8. Gustafson, *Ethics from a Theocentric Perspective*, 1:237.
9. Ibid., 238.

not of our creation, we are continually sustained by natural, social, and historical processes to which we give tacit consent."[10]

The symbol of God as Judge stems from the common senses of remorse for transgressing the limits discernible in the ordering of the world and of failing in our sense of obligation to work to sustain the natural and social ordering that others depend upon. As such, the symbol of Judge relates to the human sense of omitting to act in accordance with discernible obligations as well as committing acts that transgress limits. Discernment of limits and possibilities in relation to the ordering of the world is not always obvious. An overly strong emphasis on God's providential determination of all events, particular and general, for Gustafson, is incongruent with the sense of remorse and the moral necessity of upholding the possibility of moral accountability. In other words, it is in tension with human freedom. In a partial resolution to this, Gustafson emphasizes that alongside of his understanding of God as Judge is his view that God's activity in the world is a process of dynamic ordering rather than an instantiation of an immutable order. This divine ordering of the world is one into which humans can and do intervene.

Gustafson is here arguing that the affirmation of divine sovereignty must leave space for human freedom and possibility if the sense of remorse is universal and if the symbol of God as Judge is to be meaningful. Gustafson writes of this that, "since we are not given a blueprint of an eternal divine order, and are not given precise commands that are to be obeyed in all circumstances, the divine ordering is discovered in the processes of human experience."[11] So, Gustafson claims that the meaning of the symbol of God as Judge can be derived from the common human sense of remorse, and that both the symbol and the sense to which it correlates are grounded in the various constructive and destructive ways that humans intervene in the world.

The symbol of God as Redeemer is also rooted in human experience. While the sense of remorse is universal, stemming from the fact that humans often transgress the limits of the world's ordering and fail to conform to our sense of obligations, Gustafson also claims that the senses of possibility and direction are universal. These universal senses

10. Ibid., 239.
11. Ibid., 245.

relate to the experience of liberation, theologically construed by the Redeemer symbol. The sense of dependence and the experience of being enmeshed within patterns and processes beyond human control can yield a sense of bondage. But the senses of possibility and direction and the correlated symbol of God as Redeemer derive from the experience that to whatever degree humans exist in a world that is not of our own making, and to whatever degree humans fail to meet our sense of obligations, there yet remains within us a capacity to self-correct, to intervene, alter, and improve the world and ourselves.

It is somewhat intriguing that Gustafson does not articulate a Christology in his account of God the Redeemer. Instead, he includes it in the section of the first volume of *Ethics from a Theocentric Perspective* that follows his account of the relationship between theology and the sciences, which I interpreted in chapter two, and allots only five pages to it.[12] But the relatively few pages he gives to a discussion of his Christology proper does not mean that Gustafson does not grant a prominent place to Jesus in his broader theological vision. Gustafson's theology is radically theocentric rather than christocentric. In keeping with Reinhold Niebuhr's understanding of this, Gustafson embraces a christomorphism rather than a christocentrism.[13] Christ is built into the structure of his theological vision and profoundly *informs* his theocentrism.

Saying this does not yet explain *how* Christ informs Gustafson's theocentrism. Understanding this entails consideration of what Gustafson means by saying that "Christology is the most critical doctrinal issue for any Christian theology." I take him to mean at least two things here. First, as he puts this toward the end of his discussion of his

---

12. Ibid., 275. McCormick goes as far as to suggest that Gustafson's Christology "appears almost as an afterthought. Indeed the major directions and themes of *Ethics from a Theocentric Perspective* would stand without these [five] pages." See McCormick, "Gustafson's God," 58. As I will argue soon, McCormick's judgment here does not appreciate the possibilities of Gustafson's christomorphism and assumes univocity in the history of christologies.

13. Richard R. Niebuhr provides a helpful, basic distinction between christocentrism and christomorphism. While christocentric theologies derive all formations of doctrine from a doctrine of Christ, christomorphism means that "the redeemer is only one among a plurality of objects of theological knowledge, but at the same time he is paramount and central as the agent who reforms and shapes anew [all] the Christian's relations to God, the world and himself." See Niebuhr, *Schleiermacher on Christ and Religion*, 212.

Christology, "Theology is the noun, Christian is the modifier." To the extent then that a theology is presented as a *Christian* theology, some account of the meaning, significance and nature of Christ is obviously a critical issue. Second, Christology is a critical issue for Christian theology since, "Every effort to formulate a coherent Christology is selective of the biblical materials, and is determined in part by the issues that theologians and the churches face in particular times and places."[14]

The effort to articulate an orthodox Christology, Gustafson reminds, took several centuries (the formulation of one person and two natures was not fixed until the Council of Chalcedon in 451, and even this creedal formulation does not determine precisely how these natures should be related), was influenced by issues that were not always doctrinal (for example, the complex political situation within the Roman Empire during this time, the ecclesiastical rivalry between Alexandria and Antioch, the terminological difficulty of working between and among Jewish and Hellenistic philosophical and religious symbols), and was based upon New Testament texts that disclose a variety of ways of conceiving the significance and nature of Christ.[15]

In reference then to his affective account of religion and his preference for the narrative disclosures of the life and teachings of Jesus in the Synoptics, Gustafson's Christology is summed up in the claim that, "Jesus incarnates theocentric piety and fidelity."[16] Among the various Christological formulations, Gustafson's leans very strongly toward the subjectivist moral-influence type represented by the medieval thinker Peter Abelard.[17] This type of Christology emphasizes the transforma-

14. Gustafson, *Ethics from a Theocentric Perspective*, 1:275.

15. Gustafson's own textual selectivity emphasizes the narratives of the synoptic gospels over the Gospel of John, Revelation, and themes from the Pauline corpus. This preference for the Synoptics, he explains, stems from a judgment about the priority of narrative over abstract language, a rejection of biblicistic revelationalism, and suspicion of conceptions of the "preexistent Christ" in sources such as John, Colossians, and Ephesians. For both the quotation and the account of his textual preferences see *Ethics from a Theocentric Perspective*, 1:275.

16. Ibid., 276.

17. This is not at all to say that this understanding of the nature and significance of Christ begins with Abelard. In the early third century, for example, Clement of Alexandria wrote, "For Christ came down, for this he assumed human nature, for this he willingly endured the sufferings of humanity, that by being reduced to the measure of our weakness, he might raise us to the measure of his power" (*Quis div.* 37 [LCL]). At least part of Clement's intent here, as I understand it, is to emphasize the morally and

tive significance of Christ within subjective human consciousness. Christ is understood as a demonstration of God's love for the world, a supreme moral exemplar. This contrasts with the more objectivist type of Christology, associated for example with Anselm of Canterbury.[18]

In Gustafson's Christology, the gospels reveal Jesus as one who embodies the piety appropriate to the powers that bear down upon and sustain life, and "make clear the costs of such piety and fidelity, as well as their beneficial consequences for others."[19] Gustafson suggests that consent to Jesus' powerful demonstration of the life of piety and fidelity makes it possible to speak of him as the Christ, and that such consent to him in the early church and by the figures who surrounded him during the time of his life is one of the fundamental themes in reference to which the history of Christianity has developed. In support of this, he writes, "The only good reason for claiming to be Christian is that we continue to be empowered, sustained, renewed, informed, and judged by Jesus' incarnation of theocentric piety and fidelity."[20] Common human experience in the world can be understood to resonate with the gospel accounts of Jesus' world and his response to it. This understanding vividly reflects the character of Gustafson's vision as theocentric and christomorphic rather than christocentric.[21]

---

religiously *demonstrative* theological significance of Christ. In addition to Clement, Augustine also stressed a Christology based on a view that the life and teachings of Jesus were to be understood as a demonstration of God's love meant to evoke a response of love from humanity. My point in drawing attention to this history is to underscore that, against the thinking of some of his critics, Gustafson's subjectivist Christology (Jesus as the demonstration, moral exemplar, or "incarnation" of theocentric piety and fidelity), stands significantly in line with rather than beyond the fold of a key christological trajectory within the history of Christianity.

18. The objectivist Christology of Anselm is shaped by a "satisfaction" theory of atonement, referring to the idea that human sin is a radical offense to God that needs to be compensated for prior to forgiveness. This satisfaction cannot be accomplished by any human effort but is secured by way of the suffering of Jesus Christ, the sinless God-man. This understanding of atonement is one of the features of objectivist christologies that leads to the christocentric view of the nature and significance of Christ as determinative for the surrounding doctrinal and ethical dimensions of a Christian theology.

19. Gustafson, *Ethics from a Theocentric Perspective*, 1:276.

20. Ibid., 277.

21. Chapter 6 on the normative dimensions of Gustafson's project will present continued exegesis of this theocentric role of Christ.

In sum, Gustafson's interpretation of the gospel accounts of Jesus' life leads him to construe Christ as a persuasive and transforming example of radical obedience to God and of the rigor of moral sensitivity especially to oppression and poverty. As such, Christ is an example that can "bear down on the receptive human spirit with [its] own compelling power."[22] He acknowledges that this understanding of Jesus may not meet the demands of some interpretations of theological creeds, that, to many, "it may reduce the significance of the tradition to a social-psychological function, to a way of sustaining and vivifying a memory and way of life that has more significance for the subjective side of piety than the objective knowledge of God."[23] But he responds to this by suggesting that his Christology does not at all render Jesus meaningless or ineffective. Instead, his Christology and more broadly his revised theological symbols for God, indicate the way in which historical Christian narratives and traditions "in-form" religious affectivities, simultaneously motivate and express piety, and "move us toward the faithful consenting to the powers of God that is required for moral life."[24]

Gustafson's Christology and his interpretation of the classic symbols of God are forged between traditional theological interpretations, common human experiential verifications, and what can be validated by contemporary scientific interpretations of the world. At the same time that they aim to articulate a specifically theological construction of experience, they also "express religious affectivities in response to many aspects of nature, history, culture, society, and self."[25] He argues that though his understanding of God is shaped by and revised within a particular theological tradition, the experience and senses underlying its meaningfulness are not tradition-specific.[26]

---

22. Gustafson, *Ethics from a Theocentric Perspective*, 1:249.

23. Ibid., 278.

24. Ibid.

25. Ibid., 251.

26. A helpful way to understand this is that Gustafson has articulated the God of Christian monotheism, rather than the Christian God. Particular Christian, Jewish, or Muslim inflections of monotheism are forms of piety rooted in the historical experiences of particular traditions and people. Gustafson's historicist sensibility does not at all take these inflections lightly. And yet his revised Reformed theological framework can be thought of as an attempt to excavate a non-idiomatic construal of the divine through the tradition's historical idiom. Assessment of this demands working through the tension between historicism and naturalism in Gustafson's methodology. This is I

By defining his project as theocentric, Gustafson challenges what he takes to be Christian theology's traditionally anthropomorphic tendencies—its tendencies to attribute agency to God and the anthropocentric axiological fallout of this, in short, its constriction of the moral nature of the divine to human conceptions of "moral." He claims, as explained earlier, that anthropomorphic imagery for God is incompatible with Reformed Christianity's strong emphasis on God's sovereign otherness. While agency is only attributed analogically to God, it always remains anthropologically tethered. This for Gustafson internally contradicts the priority of sovereignty within the Reformed tradition, and, as I will show further along, is a manifestation of what he describes as the basic human fault of contraction insofar as the idea of divine agency is often manipulated to underwrite axiological anthropocentrism. For these *theological* reasons, doctrinal and ethical, Gustafson rejects the anthropomorphic and anthropocentric tendencies in traditional Christian theology and presents revisions to the symbols of God as Creator, Sustainer, Judge and Redeemer, articulating the grounds for a theocentric and christomorphic rather than a christocentric theological ethics.

But since tradition-internal criteria of coherence for Gustafson do not hold priority over the more public criteria of intelligibility and persuasiveness, as chapter two argued, he also critiques anthropomorphic accounts of God with reference to common human experience and the natural sciences. According to Gustafson there is no experiential warrant for reasoning analogically from the character of human life to the nature of the ultimate source of reality. Experience indicates, and the natural sciences confirm, that the human species has no primacy within either the long history or the present moment of nature's patterns and processes. As Gustafson well puts this, though humans may be the only "measurers" of the world there is ample evidence that humanity is not the world's "measure." The following section looks to Gustafson's fundamental ideas about the world and provides a crucial context for coming to terms with the ethical and theological implications of this claim.

---

think crucial to an understanding of Gustafson's theology. Some critics, as I will review in chapters to follow, critique Gustafson for not presenting a "Christian" God. But this critique stems in no small degree from a misunderstanding of the theocentric rather than christocentric nature of Gustafson's project.

World

Gustafson's understanding of the patterns and processes of nature, history, culture, and society—which for the sake of simplicity I refer to as his fundamental idea of the "world"—is very difficult to tease apart from his understanding of human being. The human for Gustafson is a thoroughly relational being, existing within a world comprised of a matrix of object relations, which Gustafson isolates as natural, historical, cultural, social, and the self's experience of itself. The isolation of these categories is artificial and merely heuristic; in actual experience they always overlap.

Experience of nature, for example, is never historically unmediated. And to the extent that the self is fundamentally relational, the experience of selves as natural, biological beings is always socially and culturally charged. One never experiences "nature" as a whole or in the abstract, but only through particular natural events, processes, or objects. It is almost too facile to point out that humans are deeply dependent on natural processes and other forms of life for their wellbeing. Without sunlight and rain, nothing could live. Without bodies with the capacity to reproduce, no human would come to be. Without food to eat, without rest, without the shelter provided by natural resources, without a hospitable climate, nothing could live. Clearly, human experience in nature can evoke a sense of dependence on powers beyond human control. And this experience and sense of dependence on nature, and the fact that nature continues to provide what is needed for the sustenance of life can evoke the sense of gratitude. The senses of dependence upon and gratitude for natural processes can evoke the sense of obligation to do what is necessary to sustain these processes and not to overtax them. But the powers of nature can also, of course, do harm. Too much sunlight burns skin, dries the soil, dehydrates bodies. Nature often seems indifferent to human needs. Tornadoes and tsunamis and mudslides do not detour life.

The experience of nature's threat, the fact that nature can cause harm, has throughout history motivated humans to invent ways of managing nature. Though no one can argue against the fact that human ingenuity has in many cases improved the quality of human life, the motivation to manage nature can obviously exceed itself and become exploitative. That such has occurred, does occur, and will continue to occur can evoke in some the sense of remorse for having transgressed limits. Though human life in many ways is dependent on nature, human life is not determined by nature. The transgression of natural limits is chosen. That nature can be altered, improved as well as harmed, evokes the sense of possibility as well as remorse. Nature itself and experience in nature provide "the conditions of possibility for developing our most distinctive human capacities."[27] In other words, the experience of nature is part of what funds the basic human sense of direction, of cultivating and aiming for the fulfillment of unique human capacities. In sum, nature for Gustafson is an ordering of patterns and processes that both transcend and ground life, human and otherwise.

Experience within culture and history interpenetrates with experience of nature. The sense of human possibility that nature can evoke is also culturally evoked. The sense of dependence on nature can be culturally acknowledged or denied. The majority of the global population now lives in urban environments in which the experience of nature is so radically mediated that dependence can easily be forgotten. Human dependence on other animal and plant life is thickly concealed when food is purchased prepackaged and often pre-prepared at grocery stores. The fact that a significant portion of the human population could not survive without the mediations of culture signals the degree to which experience of culture can evoke a sense of dependence. Cultural artifacts and technologies, symbolic systems, and the various sciences and arts generate a "second nature," for many the primary world in which life is experienced and by way of which the senses of dependence, gratitude, and possibility are evoked.

History for Gustafson encompasses those patterns and processes initiated primarily by human freedom, and historical experience always mediates experience of nature and culture. Historical events then are human produced. And yet many humanly produced historical events have consequences for nature. Global climate change is literally an historical and culturally *fueled* event with natural effects. So the world of nature and the worlds of history and culture overlap. As with nature, historical and cultural patterns and processes transcend and ground human experience.

---

27. Ibid., 210.

Society and the self, too, are among the basic domains of human experience of the world. Experiences within social institutions and of the self are always tethered to other dimensions of experience. The self's relation to its body, for example, is one of the most basic experiences of nature. We depend on our bodies, can be grateful for them, can transgress their limits and feel remorse for doing so. The self's experience of the body is also always influenced by cultural ideas and images of it, by historical and scientific and medical developments that alter the conditions and possibilities of embodied life. The various forms and institutions of social life, of course, also reflect experience in nature, culture, and history. The family, for example, is partly a natural, biological unit and partly a cultural and historical artifact. Forms of government and economic orders, too, are social experiences basic to the human experience of the world. Governments and economies can both sustain and constrain the possibilities and directionality of human life and influence and are influenced by each of the other spheres of experience.

Together, Gustafson's analyses of the experience of the world within the domains of nature, culture, history, society, and the self, and the senses they evoke, shapes and is shaped by a deep vision of interdependence. As shown, it is difficult to disentangle the basic spheres of experience, and this is part of Gustafson's point. Experience and its modes are multifaceted; the self as relational exists inextricably bound up within the interpenetrating domains of experience. Experience in each of these domains can be correlated to the various basic senses. Focus on the basic senses and the different objects and experiences evoking them highlight the webbed character of experience and the interrelationship of all things to one another. The world is a field of interdependent patterns and processes.

Gustafson's emphasis on the world's intermingling patterns and processes is roughly amenable to ecological descriptions of the interdependence of natural systems, but he argues against conflating a description of interdependence and a norm of harmonious equilibrium.[28] Natural processes are dynamic, always shifting, and in constant flux, while equilibrium suggests a stability of relationships, an idealized vision of natural systems. Until relatively recently, ecological science was guided by the idea that natural systems tend toward and eventually

---

28. See Gustafson, *A Sense of the Divine*, 1–20.

reach a stable climax. Ecosystems, when undisturbed by human interventions, were thought to move through stages of succession toward equilibrium. But the consensus now is that change is one of the few ecosystemic constants.[29] While ecological science has not at all dispensed with the view that natural systems are constituted by interdependent relationships, it has radically challenged the position that natural systems tend toward equilibrium.

Thus Gustafson's most basic descriptive claims about the world are that it is interdependent and dynamic. But he also acknowledges that his view of interdependent flux generates a practical problematic. If flux rather than equilibrium is the norm, one cannot look directly to nature for moral guidance of human interventions into it. If change is the standard within nature, there seems to be no unambiguous normative reference point in nature around which to organize the human moral relation to nature. And so Gustafson's fuller vision of the world, inclusive of history as well as nature, is a vision of morally ambiguous, interdependent patterns and processes, divinely governed.

Human Being

Central to any description of human beings is some account of the degree to which humans are capable of freely directing their own lives, and what Gustafson refers to as varying accounts of the "margin of freedom" characterizes different moral anthropologies. On one end of the spectrum of freedom stand "the extreme claims of existentialism," and on the other end, "highly deterministic views based on psychoanalysis, the social sciences, or brain physiology."[30] With respect to the pole that emphasizes radical freedom, it is held that humans are not captive to natural instincts or to our biological natures in the same way that other animals are. In fact, what such views typically take to be decisive about human nature is the human capacity to transcend nature. But on the deterministic end of the spectrum, freedom is radically qualified by a description of humans situated within and conditioned by the same forces or laws operative throughout the rest of nature. Such accounts tend toward determinism in contrast to freedom.

---

29. See, for one ecological scientific interpretation of this, Botkin, *Discordant Harmonies*.

30. Gustafson, *Ethics from a Theocentric Perspective*, 1:281.

For thinkers like Gustafson and Jonas, for whom articulating the relation between humans and the rest of nature is a central task, neither option of extreme freedom or determinism provides an adequate description of human nature. As I have been arguing, it is crucial ethically to balance human continuity with and difference from nature, to provide an account of human life that neither denies the human place within nature nor unique human capacities. If a description of human being is to be adequate, then it must articulate human distinctiveness without overriding biological continuity with the rest of life. And if a prescription for how rightly to live among other humans and the rest of nature is to be morally sufficient, it must include a view of humanity's morally consequential, biologically distinct capacities. The demand for such an anthropological balance is not solely that good science requires it. The demand is ethical as well, as I suggested in my introductory chapter. While the previous chapter demonstrated the way in which Jonas negotiated the Scylla and Charybdis of human continuity and difference from nature with his anthropology of responsibility, it is now time to examine Gustafson's moral anthropology of participation.

Gustafson's moral anthropology emerges initially through a focus on themes and metaphors of human sociality. Any model of human moral nature, he acknowledges, is influenced by images of human social relations. Gustafson notes two historically dominant social metaphors, the organic and contractual, both of which falsify the human condition and experience. Whereas an organic social metaphor "excessively highlights the processes of continuous mutual determination between persons, between groups, and in some instances, as in the extreme sociobiological views, between human beings and the rest of nature," a contractual social metaphor highlights and gives primacy to individuals.[31] In contrast to the organic metaphor, the contractual one grants greater efficacy to human individuals in the determination of social events and experiences and tends to stress individual human discontinuity from natural processes. In contrast to the contractual metaphor, the organic one describes human individuals more as effects than as causes in social and natural processes. The contractual metaphor overplays individual autonomy while the organic one downplays it. Both accounts of sociality include significant insights, but to the degree that they tend to

---

31. Ibid., 292.

highlight particular dimensions of human experience over others they falsify a more capacious rendering of it. In addition, these metaphors lead to different and equally inadequate accounts of how to negotiate ethical conflicts between individuals and social and natural communities. While the organic metaphor grants moral priority to communities over individuals, the contractual metaphor grants the reverse.

In contrast to the contractual and organic models, Gustafson argues for an interactionist one and for a moral anthropology organized around the metaphor of participation. As with the organic and contractual models, the interactionist one includes an account of the human relation to nature as well as to the social order. Along with the organic model, it affirms that humans are socially and naturally conditioned. Along with the contractual model, the interactionist one affirms the causal efficacy of individual humans and of the human social order within nature, but without overemphasizing this efficacy. Against both models, the interactionist model suggests a less unidirectional view of causality. Neither individuals nor communities have causal priority. Along these lines, Gustafson writes that, "Multicausality . . . must be taken into account in understanding individual, social, historical, and cultural developments."[32] Multicausal interactionism, in which influence flows back and forth between individuals and communities or systems, is the social model that best supports a description of the multidimensional experiences of humans as participants within social and natural patterns and processes. Normatively, the interactionist model does not provide an abstract calculus for determining moral priority in conflict cases. Instead, it is a description of social relations leading to the view that critical moral judgments and priority valuations are inherently ambiguous and often tragic. Such judgments and evaluations can only be circumstantially determined; they cannot be abstractly pre-determined but must be negotiated with reference to specific contexts.

Gustafson's argument for the adequacy of this model of social relations illuminates his participatory anthropology and its emphases on the inherently valuation character of human life and the importance of moral accountability and sin. Citing the British moral philosopher Mary Midgley, Gustafson argues that the proper question to ask about humans is not what distinguishes humans *from* other forms of life,

---

32. Ibid., 293.

but what distinguishes humans *among* them.[33] Understanding human moral life needs to be approached through an account of the whole nature of humans within the natural and social worlds. Several moral anthropological insights are generated when the meaning of human being is approached in this way.

First, such an approach acknowledges interdependence in addition to human dependence, prioritized in Gustafson's account of the natural senses. Humans are brought forth through biological processes and are shaped as persons through the interaction of those processes with cultural and social ones. Such dependence is a fact of existence for the human species. Though this recognition of dependence "requires acknowledgment of limitations and of the possibilities for human initiative and development," it "does not imply enslavement."[34] While underemphasizing dependence leads to an emphasis on human autonomy that can neglect biological conditions, overemphasizing it can lead to a morally dangerous emphasis on passivity.[35] For humans, like all other living organisms, are not only dependent on patterns and processes beyond their control but also exist in interdependent relation with them as they continually shape them in the aim of continued existence.

So natural and social interdependence is affirmed by Gustafson's interactionist model of human social and natural relations. While individual humans are brought forth by biological processes they do not control and are brought into cultural or social patterns that are not of their own making, humans have the capacity positively to direct their lives. Biology and culture are both inherited, but neither is fully determinative. Human capacities are efficacious, capable of altering the social and natural orders, capable, to some extent, of predicting how such alterations may impact the future. Some settings in which humans find themselves are more susceptible to alteration than others. Human efficacy and the capacity to predict the interactive effects of action are always relative to the spheres in which that efficacy operates. In sum, an interactionist model for human social relations reflects and is reflected by a moral anthropology in which humans are construed as natural historical participants.

---

33. Ibid., 282. Gustafson quotes here from Midgley, *Beast and Man*, 203.
34. Gustafson, *Ethics from a Theocentric Perspective*, 1:282.
35. Ibid., 283.

This connection allows Gustafson to affirm a second key insight in his fundamental idea of human being, the significance of desire and valuation: "Man is a *valuing* animal; as such, man shares a great deal with other animals."[36] Gustafson is affirming here that commonality rather than difference between humans and other animals is an important starting point for moral anthropology. He affirms Jonas's insight that whatever species-characteristic humans have are legacies of natural processes as well as cultural ones. Most of the history of philosophical and religious thinking has deemed that a focus on human discontinuity with other animals is necessary to a proper estimation of human nature. Gustafson resists this by arguing that any proper treatment of human being should at least also include a view of human kinship with other animal species.[37]

The human is a valuing animal. Valuation is linked to biologically rooted needs and feelings, to propensities to desire and want. Other animals too value. They purposively aim for objects they desire and flee objects they fear. Purposive activity, at bottom, is activity directed toward objects valued on the basis of need, desires, and feelings, the most basic of which are biologically structured.[38] Values arise in relation to purposes and the conflicting motives for actions toward desires and needs. Valuation is part of animal and thus human nature. This is not to deny, however, that reason is also an essential aspect of human nature. The prioritization of certain values over others and deliberation about means to attain purposes is an effect of the application of reason to the deep roots of human animal desire. According to Gustafson, "We are distinguished [from other animals] . . . by our capacities to examine critically various objects of our desires, ends of our motives, and objects of our valuations."[39] Human reason is exercised not apart from biological propensities but along with them. And yet it differs from the purposive means-end calculations of other animals. For according to Gustafson, human valuation is reflexively aware of the relations of parts and wholes. In other words, desire and valuation are contextualized, or framed, by a relative horizon of concern. Valuation is inescapably

---

36. Ibid.

37. Ibid., 282.

38. Gustafson seems to limit purposiveness to animal life, whereas Jonas reads it all through life.

39. Gustafson, *Ethics from a Theocentric Perspective*, 1:286.

relational. As my discussion below of Gustafson's understanding of the basic human fault will explain, though human valuation is generally relationally contextualized, the circumference of the context tends to be anthropocentrically construed. The normative dimension of his theocentric ethics, building upon this account of the human as valuational, will argue for an enlargement of the context of human moral concern and religious piety.

The historical moral anthropological prioritization of reason over feeling and desire tends to downplay the biological continuity of humans with other animal life that Gustafson is emphasizing here. But arguing against the primacy of reason does not mean either that humans do not reason in choosing and ordering values or that such reasoning is not crucial to a full account of human being. But it does focus attention on the relations obtaining between reason, desire, and feeling and stresses biological continuity rather than discontinuity. Affirming human being as valuational primarily dissolves the strict divide between feeling and reason, and accords as well with Gustafson's earlier characterization of religion as largely affective. As Gustafson writes, "To affirm valuing as the central descriptive term for man . . . is to claim that affectivity is a principal feature of *all* human activity."[40] As stated before, "to think about [reason and feeling] as if they were separate characters in a drama of deep conflict, is to falsify the nature of man."[41]

Having said that humans are basically valuational, and by way of that claim emphasizing biological continuity, Gustafson specifies his understanding of freedom and agency. As mentioned above, two problems are crucially related to this, descriptive and prescriptive. On one hand, descriptive programs in the modern sciences stressing nature as law-governed and deterministic problematize the notion of freedom. This was the problem Kant attempted to overcome through his differentiation of the phenomenal realm of nature and the noumenal realm of freedom. For Kant, freedom is a noumenal idea, a necessary postulate of practical reason if human action is to be effective in the world. The second problem of freedom is prescriptive. If the idea of freedom is undertheorized or unfulfilled, then the very concept of morality cannot be sustained. The possibility of holding oneself and others mor-

---

40. Ibid., 287 (italics mine).
41. Ibid.

ally accountable for actions depends on the capacity freely to choose particular courses of action. If one is not free, intention and choice have no meaning, and if intention and choice have no meaning, moral prescriptions and proscriptions have no real bearing. Gustafson's response to these two challenges, similar to Jonas's, is to argue for a notion of freedom partly built upon biological theory.

Humans are biological creatures embedded in natural and social processes that simultaneously generate, sustain, and limit freedom. As should be clear by now, this view of freedom for Gustafson does not privilege one dimension of human experience over others, such as reason, but references the whole nature of a person. Gustafson writes of this integrative vision of freedom that, "whether moral or otherwise, [our intentions and choices] draw upon and give focus to our biological natures."[42] And yet at the same time, human social, cultural, and historical circumstances also bear on freedom. Such circumstances both limit and make possible freedom's particularized expressions. Habituation, socialization, and formation within cultural traditions and their beliefs and practices condition "the limits and possibilities of action at a particular time."[43] Gustafson emphasizes here that though natural and social circumstances condition human freedom, they do not undermine moral accountability. Natural capacities and propensities may be given, and social, cultural and historical circumstances may shape them, but humans can freely choose to nourish or neglect them, intentionally to increase or decrease possibilities, to expand or contract them. This means according to Gustafson that we "are accountable for the ways in which we bring our 'natural' capacities to a focus of choice and action, for the assessment of our interrelations with other persons and other things in determining how we will exercise our powers, and for the understanding of the circumstances of our action."[44] This claim affirms the claim earlier referenced that a multicausal account of human efficacy is necessary to the multidimensional character of human experience.

Thus, for Gustafson, human experience is multidimensional and relational because humans are socially interactive animals. The multi-

---

42. Ibid., 290.
43. Ibid.
44. Ibid., 291.

dimensional nature of human experience demands a complex account of the various conditions that fund it. Accounts of human moral nature are strengthened rather than weakened when they include the biological conditions or structures of human relationality. Gustafson names "participation" as the root moral anthropological metaphor that evokes these ideas. Humans are biological and cultural creatures that participate in a field of patterns and processes, in relation to a power or powers that operate around, through, and beneath them. Participation with and relation to powers or a power beyond human control gives rise to what Gustafson views as a natural sense of accountability and leads to the theological dimension of his moral anthropology.

As indicated earlier, Gustafson argues that the sense of accountability is universal: "the human experience of accountability for moral wrongs is present in all persons who accept the fact that they are agents and that they are responsible to some extent for their choices and their actions."[45] Though universal, accountability is conceptualized in multiple ways by various philosophical and religious traditions. Particular views of the origins of accountability and its extent are tradition-shaped. But all particular views of accountability are built upon what Gustafson considers to be four interrelated facets of the universal human sense of fault—first, the experience of misplaced trust or loyalty; second, misdirected valuations; third, faulty perception and understanding of the relations of things to one another; and fourth, the experience of unmet obligations.[46] The Christian theological doctrine of sin and other tradition-specific accounts of fault arise according to Gustafson from these four universal experiences. At their core is the experience and consequence of "a disordering of proper human relationships to other persons and to the world around us and each involves a somewhat disordered relation to the powers that sustain life and bear down upon it."[47] As Gustafson understands this, the platform for the expression of each of these faults is an improper sense of human finitude.

The term Gustafson proposes for understanding this network of fault is "contraction." Contraction draws from and reinterprets the Christian theological doctrine of the "fall." Though it has a place in the

45. Ibid., 293.
46. Ibid., 294.
47. Ibid.

tradition, Gustafson's understanding of contraction resists the implication in the idea of the fall that humans ever existed without this condition. He writes, "We know too much about how our species developed biologically and culturally even to dream of sustaining the notion that there was once the purity of vision ... from which we have fallen."[48] The contraction of the human spirit is embedded in finite human nature. Though contraction is thus to some degree unavoidable, humans can be held accountable for too easily succumbing to it or for not taking up the possibilities of correcting it. The possibilities of correction will be discussed in the normative chapter; for now it remains to describe the four universal experiences that finitude and contraction produce.

First, the experience of misplaced trust can be configured in religious terms as idolatry. The experience and sense of misplaced trust, of wrongful loyalty, stems from placing too much confidence in objects that are not sufficient to total trust. Idolatry is a religious term for the universal experience that humans have of prioritizing objects in the quest for meaning and purpose that are not adequate to the quality of meaning and purpose that is sought. Along with his mentor H. Richard Niebuhr, Gustafson is here affirming that every human agent organizes his or her life around some center of value or values.[49] When false or inadequate to the projects of the moral life, when insufficient to the multiple dimensions of experience, this center of value or locus of moral gravity produces moral distortion. In essence, the fault of misplaced trust is the fault of placing too much confidence in one object in our experience over the others, an object unworthy of ultimate loyalty. Placing ultimate loyalty on such limited objects produces moral distortion insofar as it yields a misconstrual of the proper relations of things to one another.

The experience of misplaced trust is related to misdirected valuation. Humans can never foresee with perfect clarity which objects are worthy of ultimate loyalty and which are not. As has been shown, humans for Gustafson are basically valuational creatures. But in contrast to the means-ends valuations of other animals, and as a result of naturalistically grounded but distinctive human cognitive, affective, and moral capacities, Gustafson understands human valuation according to

---

48. Ibid., 305.

49. On this, see especially H. Richard Niebuhr's classic essay, "The Center of Value."

a part-whole dynamic. The fault of misplaced valuations is not rooted in desire itself, but in the contraction of desire. For example, while health is integral to human flourishing, and thus a right object of value and something good to be desired, when the desire for health overrides the value of the whole complexity of human flourishing, when it is in other words excessively valued, it produces moral distortion. The problem with disproportionate desires, argues Gustafson, is learned experientially. For example, with the case of a fitness fanatic, in time she may learn how such fanaticism corrodes other goods in life, such as family, or the life of the mind. This second universal fault, then, is experienced as the tendency to desire and value certain objects or ends in excess of what is proper to them. As complex, multidimensional beings, humans depend on and desire many things. Misdirected valuations result from a failure to sort or organize desires in right proportion to one another, in relation to some proper object of ultimate trust. In the language of Augustine, this is the fault of disordered love.

These two universal faults relate to the third. Misplaced trust and misdirected valuation are produced by and sustain the fault of reason, of improperly seeing and understanding the part-whole relation of things to one another. Gustafson describes this as principally a rational-interpretive problem. He writes, "The fault of rationality is not so much a matter of errors in logic as it is in misconstruing the realm of reality that engages us; it is a matter of the wrong depiction and interpretation of the particular 'world' that attracts our attention and that evokes our activity."[50] This hermeneutical view of the problem of rationality relates to Gustafson's methodological emphasis on the construal of circumstances described in chapter two. One of the requisites of right moral action is a reasonably accurate account of circumstances. Failure to provide a minimally reasonable interpretation of circumstances is a moral fault.

Gustafson is here making an important point that stakes a claim with respect to his Reformed theological tradition. Against the tradition's prevalent tendency to describe human reason as "totally corrupt," Gustafson argues that while the fault of interpretive-reasoning may be fundamental, it is not something for which in principle we are morally culpable. Since every human is situated within a cultural and historical

---

50. Gustafson, *Ethics from a Theocentric Perspective*, 1:300.

framework, and since cognitive capacities are simultaneously enabled and limited by their biological underpinnings, every person's interpretation of the relations of things to one another is perspectival. Perfectly clear and certain ideas of things are never fully possible. Perspectivalism and the limited accounts of reality it produces are inescapable and universal. For Gustafson, "We cannot be held accountable for our finitude, which is part of our nature."[51] The incompleteness of interpretive-reasoning is a universal feature of human nature because humans are finite, and this nature cannot be held against us.

However, though finitude is not something that can be escaped, humans can be held accountable for failures "to 'see' certain aspects of the world to which we are attentive, our failures to take into account relevant information and explanations, our refusals to be corrected in the light of substantial evidences and persuasive arguments."[52] While there are many things humans cannot hope to know with finality or completeness, there are many things that can and ought to be known that for various reasons are resisted. Accountability for such rational failures is linked to the fact that reasoning is always, for Gustafson, related to human desires and interests. Desire and interest are not in themselves the problem. Humans cannot help desiring and being interested, and much of the good that humans can do depends on desire and interest. Without a desire for understanding or interest in reality, for example, no work of human knowledge, creativity, or invention could get off the ground. Instead, accountability for the fault of reason derives from a culpable narrowness of mind produced by the other two universal faults, misdirected valuations and misplaced loyalties.

The fourth universal experience of fault is the experience of disobedience. Disobedience stems from false desires and interests born from and sustained by the other three basic faults. While specific religious and philosophical traditions conceive of disobedience in different ways, with respect for example to failure to obey the will of God, the laws of nations, or universal moral principles, Gustafson argues that such specific conceptions relate back to universal human experiences. The various codifications of obedience derive from the negative experience of violating basic and natural principles. Gustafson affirms, "there is

---

51. Ibid., 301.
52. Ibid.

evidence for sin [here discussed in the form of disobedience] from the history of mankind that needs no biblical terminology or authorization to be perceived."[53] What the Christian theological traditions add to the universal evidence of disobedience is the idea that it is not simply about the breaking of rules but about the total moral character of agents.

Theological insight into the root causes of disobedience in misplaced trust, misdirected desire, and faulty reasoning confirms for Gustafson that the underlying and universal fault of the human condition is best described as contraction. Idolatry is a contraction of human trust and loyalty. Misdirected valuations are a contraction of desire, based on the contraction of trust and loyalty. Faulty perception of the relation among things is a contraction of vision, related to misplaced trust and misdirected valuation. Disobedience is a contraction of moral interests. If the basic human fault is contraction, Gustafson argues, the correction needs to be understood as "enlargement." This will be further discussed in chapter six. For now it remains to consolidate what the foregoing treatments of Gustafson's fundamental ideas indicate with respect to his moral anthropology and to consider this in critical contrast to Jonas's.

Conclusion

Together, Gustafson's revised Reformed theological doctrine of God and his description of the world reflect and reinforce his moral anthropology of participation. Emphasizing human experiential inclusion within and dependence upon a matrix of natural, historical, cultural, and social patterns and processes, supported by a theological vision of the power ultimately bearing down upon and sustaining life, Gustafson's participatory anthropology contrasts in some important ways with Jonas's anthropology of responsibility. Critical interpretation of these contrasts will move this work toward its final part on the normative dimensions of each thinker's ethics and the task of consolidating an outline of a theological moral anthropology of responsible participation.

The first major point of difference concerns the ontology of the moral agent implied by each thinker's metaphor. As argued in chapter three, Jonas's anthropological metaphor reflects a relatively idiocentric ontology treating the human as an individual organism in a basically

53. Ibid., 304

agonistic relation to the environment. This is related, as the chapter on his hermeneutics argued, to the existentialist framework through which he undertakes his reading of biological theory. Unique to the human among other forms of life, for Jonas, is the degree of freedom and the magnitude of power that result from the increased mediacy in the human self- and world-relation. The role here of mediacy in Jonas's anthropology reflects the degree to which his normative theory highlights human discontinuity with the larger natural world. Comparatively, Gustafson's metaphor reflects a more relational ontology, emphasizing continuity over discontinuity. The human is a relationally interactive, morally reflexive social organism dependent upon, empowered by, and exerting considerable but not totalizing influence on the ordering of nature and history that are ultimately borne down upon and governed by God.

Each of these agential ontologies suggested by my interpretations of each thinker's anthropological metaphor contain crucial insights. But these insights become more constructive when drawn together. Jonas's anthropology accurately describes and will be shown in the next chapter to provide an important foundation for him morally to build upon his account of the responsibilities of human distinctiveness in relation to nature. Likewise, Gustafson's anthropology accurately describes and, as will be shown in chapter six, morally builds upon his emphasis on human participation within the divine ordering of the world's patterns and processes. Both of these anthropologies, I hold, bring to light crucial aspects of the human moral relation to nature. As both thinkers attempt to show, humans exist in a simultaneously continuous and discontinuous relation to the natural world. And yet the dialectical character of this is undertheorized to the extent that their anthropological metaphors subtly emphasize one aspect of the dialectic over the other. The anthropology of responsible participation this book points toward attempts to reflect the dialectical character of the human moral relation to nature more fully.

A second contrast evident in Jonas's and Gustafson's moral anthropological metaphors, related to the first, concerns their different accounts of practical reason. Though more complete treatment of this will be developed in the coming chapters, it is helpful to point forward to that discussion by relating here what their fundamental ideas suggest about it. Practical reasoning for Jonas is a more rationalistic enterprise than it is for Gustafson. Jonas reasons to the imperative of responsibility, and commends it, on what he will argue is the intuitively rational principle that value is ontologically grounded. Embedded within Jonas's anthropology of responsibility, then, is an account of practical reasoning as a principally cognitive enterprise, a matter of providing a rationally persuasive norm to guide human power.

In contrast, Gustafson's metaphor of participation shapes and is shaped by a primarily affective account of practical reasoning. This account of practical reason for Gustafson reflects the integration of intellect, feeling, desire and valuation in his anthropology of participation. This chapter has shown that Gustafson's anthropology and the theory of value it indicates are both thoroughly relational. Practical reasoning is the deliberative, simultaneously cognitive and affective, fundamentally hermeneutical enterprise of organizing one's experience of the whole of the world's patterns and processes and one's motives in the aim of relating rightly to things. The ultimate standard of a "right relation" for Gustafson, as chapter six will bring into more critical view, is that which is appropriate to a thing's relation to God, a criterion discerned within the conditions of finitude regarding what the world's patterns and processes ultimately indicate. Practical reason for Gustafson, then, is not deduction from an intuited rational principle to what it entails for action, nor a direct movement from theological indicative to imperative, but an unavoidably ambiguous interpretive enterprise engaging the affective, cognitive and valuational, capacities of the agent. The specific content granted to the formal criterion of relating rightly to things is not predetermined but arises through and is constituted by the multidimensional participant's affective exercise of practical reasoning. Practical reason for Gustafson is, in sum, the participatory task of moving from the whole matrix of relatively disorganized experience to a deliberative organization that yields reflective action in accordance with a discerned sense for what is entailed in relating rightly to the world's ultimate ordering power, God.

These different understandings of practical reason are more difficult to join than their ontologies of the moral agent. But they need not be fully integrated. As I will discuss in my next chapter and in my conclusion, the strong emphasis on knowledge in Jonas's account of practical reasoning informs a highly problematic political prescription. While he is right to suggest that knowledge has become a new and primary duty in a time of radical power, his emphasis on this leads him to neglect the significance and, indeed, the necessity of the moral formation of responsibility. For responsibility is not motivated by reason alone, and as he himself suggests, there are many reasons to think that humans will not be rationally persuaded to the life of responsibility. Gustafson's relational, affective, hermeneutical account of practical reasoning, and his understanding of the importance of communities of moral formation, I will suggest, provides an important complement to Jonas on this score. But this will be explored more fully in the coming chapters.

In sum, both Jonas's and Gustafson's moral anthropological metaphors influence different ways of seeing and sensing human individual and community life, of the moral limits, possibilities, and obligations of the human species in relation to nature. They illumine varying ontologies of the moral agent and accounts of practical reason. I argue that the full dialectic of human moral experience is better represented by an anthropology that draws together Jonas's and Gustafson's ontologies of the agent, and that Gustafson's account of practical reasoning provides a crucial corrective to Jonas's, especially on the matter of moral formation. But there is a third basic difference between Jonas and Gustafson that overshadows these, which, of course, is that while Jonas considers theology a luxury of reason, Gustafson treats the divine ordering of reality as a viable, fundamental construal of the way things ultimately are. Jonas's view of theology will be considered in chapter five, in which I will entertain the possibility that, while his ethical vision may not require theological support, he himself seems to indicate that it may be strengthened by a theological component. Entertaining that possibility presupposes a more thorough inquiry into his normative claims, and thus needs to be put off here. However, since theology is by definition central to Gustafson's project, an initial critical treatment of his theological framework is appropriate here.

In the introduction to this chapter, I noted Gustafson's claim that ethical theories generally presuppose foundational commitments that either are theological or are theological equivalents. I also stated that Gustafson views all phenomena as intersections in which disciplinary traffic flows from multiple directions and holds to the position that the traffic of interdisciplinary thinking can be organized in a way that provides more meaningful interpretation of the way things ultimately are than is possible from the perspective of a single disciplinary framework. His view is that various disciplinary discourses should and can communicate to one another and that specifically theological perspectives should and can be intelligible and even persuasive to secular sensibilities.

And yet, while it is clear that the traffic is flowing in Gustafson's work, to some of his interpreters, the destination of this traffic is unsettlingly unclear. Gustafson deems that his interdisciplinary methodology is requisite to the continuing public meaningfulness and intelligibility of Christian theological discourse. And yet, as Gordon Kaufman has argued, it is not entirely clear why Gustafson holds that the concept of God, understood impersonally in reference to the power bearing down upon and sustaining life, provides a more meaningful ultimate reference to the way things are than a more general understanding of "Nature." Kaufman asks:

> Why has Gustafson presented his ethics as a specifically *theological* position? Why has he not set it out as a form of ethical naturalism? Is *God* really as important in Gustafson's scheme as he claims? Would not a naturalism developed with ecological sensitivity provide essentially the same framework . . . as Gustafson's so-called theocentrism?[54]

This is a significant critique, and one that others besides Kaufman have raised. To the extent that Gustafson's project intends to make a theological contribution to human understanding of the way things ultimately are, and from this understanding to prescribe a theocentric ethical orientation for human life, the issue obviously demands some critical attention.[55]

---

54. Further, Kaufman queries, "Given the [totally impersonal] conception of God to which he finally comes, Gustafson's unwillingness to use a straightforward ecological naturalism as the basis framework for his ethics seems even more difficult to understand. What can the introduction of the notion of God—conceived in this completely dehumanized and depersonalized way—do for Gustafson that the notion of nature cannot do as well?" For this and the quote cited above, see Kaufman, "How is God to Be Understood in a Theocentric Ethics?" 21, 27. Kaufman is not the only interpreter of Gustafson who raises these issues. I will deal with others who level similar charges, such as Stanley Hauerwas, Richard McCormack, and Lisa Sowle Cahill in ch. 6 and in my concluding chapter.

55. I engage Kaufman's critique specifically, not because I take it to be especially wrong but because I think it is representative of a fairly common misinterpretation of Gustafson.

Kaufman does not believe Gustafson can provide adequate response to his criticism. While he basically affirms Gustafson's rejection of the anthropocentric tendencies in much of traditional Christian theology, he deems that Gustafson has "overreacted to these problems" and that this overreaction leads to the difficulty of distinguishing between Gustafson's theocentrism and an ethical naturalism.[56] As I interpret him on this point, Kaufman highlights a difference between the rejection of axiological anthropocentrism and anthropomorphic theological discourse. This is a perceptive distinction. He suggests that the problems with anthropocentrism are not necessarily overcome by "simply cutting out" anthropomorphic ways of thinking and speaking about God.[57] And further, dismissing such ways of thinking and speaking theologically without fairly attending to their possible constructive insights "is why [Gustafson] is unable to give an adequate account . . . of why it is important that ethics be theocentric and not just naturalistic . . . ."[58]

According to Kaufman's reasoning here, Gustafson's critique of anthropomorphism leads him to a conception of God "that cannot be understood as significantly present in or working through those powers that have actually brought humans *qua their humanity* into being and that continue to sustain them in being."[59] The assumptions here are that anthropomorphic imagery for God is necessary to disclosure of the ways in which God works to sustain distinctively human life and that Gustafson's construal of God does not sufficiently provide such imagery.

I take issue with this. If Gustafson has perhaps "overreacted" to the problems with anthropomorphic discourse, I think it is at least as fair to say that this line of critique has not fully appreciated the subtlety of Gustafson's reasoning. According to Kaufman, Gustafson's rejection of anthropomorphism stems from an overdrawn emphasis on the activity of God within nature's patterns and processes. If nature's patterns and processes are the sole sphere for the experience of God and the dominant source for thinking and speaking of God, Kaufman implies, an impersonalistic construal of the divine that provides no advantage

---

56. Kaufman, "How is God to Be Understood in a Theocentric Ethics?" 28.
57. Ibid.
58. Ibid.
59. Ibid.

over a purely naturalistic account of reality is bound to emerge. But Gustafson has not relied solely on experience within the domain of nature's patterns and processes for his construal of God. As my exegeses of Gustafson's fundamental ideas have shown, Gustafson interprets the presence and activity of God at work in much more than nature, for example, in and through culture and history, in and through human social relations. For Gustafson, God is the ordering power that bears down upon and sustains *all* things. This is a radically theocentric vision.

In addition to this, Gustafson challenges the operative assumption in Kaufman's charge that there is something like a "general" notion of "nature." The idea of "nature" is purely abstract; for Gustafson, there is no such thing as "nature" unmediated. There are only particular experiences within particular natural contexts in particular times, and these experiences and the ideas about the natural world that arise from them are always to some extent historically and culturally influenced. My exegeses make clear that while Gustafson distinguishes among various domains of experience (i.e., natural, historical, cultural) for analytic reasons, he very deliberately interprets the ways in which they overlap.

The overlapping of the domains of experience is related to the multidimensional character of human experience. There is really no such thing as a pure experience of nature or history since natural and historical conditions interpenetrate and since human selves are natural historical beings. Given this, the common human senses in reference to which Gustafson's construal of God largely takes shape are not restricted to the experiential sphere of nature's processes. To charge Gustafson with a construal of God that is indistinguishable from nature fails to appreciate the cultural and historical contexts of experience and understanding that Gustafson draws upon.

The multiple, overlapping domains of experience that give rise to the common human senses support Gustafson's retention of the relatively personalistic symbols for God from his tradition, even in spite of his suspicion of anthropomorphic theological discourse. What Gustafson acknowledges with his retention of these symbols, precisely through his revision of them, is that his thinking, indeed all thinking, is intractably situated within a tradition of discourse that simultaneously limits and makes possible what can be said about the ultimate nature of things. Gustafson uses his tradition's symbols for God because he has been and is conditioned by his tradition's patterns of thinking and speaking, and because he deems these symbols to offer a rich perspective on the way things ultimately are, a perspective that would not be as rich without them. This reflects the deep historicism of Gustafson's sensibility. At the same time, he critically revises these symbols in light of common human experience and the insights of the sciences because his historicism is naturalistically qualified. No tradition, on Gustafson's judgment, whether theological or scientific or otherwise, is closed in upon itself.

Critiques of Gustafson such as the one here considered stem in significant measure from the fact that Gustafson not only acknowledges this understanding of the developmental, porous character of all discursive traditions—which many thinkers acknowledge—but also because he actually presents an ethical system that fully incorporates this understanding. The result is a supple perspective on the theocentric moral life that features a finely textured portrait of the tangled bank of human participation within nature and history and in relation to the ordering power of God. If there is fault to be found in this picture, I will argue, it is not in the scholarly rigor with which it is presented, but perhaps in the normative proposal to which it leads and its possible insufficiency for the grave responsibilities of human life in a time of environmental crisis. Gustafson's interdisciplinary methodology, his historicist and naturalistic sensibilities, may themselves, in light of the contemporary context, demand and aid a revision to his central normative claims. Inquiry into this possibility will be taken up in chapter six, and in keeping with my rhetorical strategy and my overall purposes, can only fully be treated once the normative dimension of Jonas's project has been interpreted.

PART THREE

Normative Dimensions

5

The Imperative of Responsibility

Introduction

ACCORDING TO JONAS, AS INTERPRETED IN CHAPTER THREE, THE HOrizon of freedom, originating primordially in metabolic processes and expanding through the variety of life's forms, reaches its pinnacle in the nature of human beings. While all living beings purposively affirm their existence metabolically, this purposiveness reaches a conscious level only through the uniquely mediated character of human existence. Humans metabolize like all other organisms. Humans are mobile, perceiving, and feeling, as with all other animals. But only humans create and use tools that are themselves worked upon and made available for future work. And by way of these tools humans gain a capacity for manipulating the world on a magnitude unknown to other forms of life.

Only humans produce images and therefore inhabit in addition to the world of nature a second, constructed, symbolic world. Though imagined, this world of images can impact the world of nature insofar as what is imagined acts on, conditions, and can reshape human perception of and activity in the natural world. The constructed second world of images thereby can alter the world of nature. By way of the world-mediation brought about by tool-use and image-making, human power over the world becomes qualitatively different from the powers of other forms of animal life.

Further, only humans are consciously aware of mortality, memorializing their lives by creating graves and rituals of death. Consciousness of death, for Jonas, leads to self-mediation, to the self's relation to an "I," and to the idealized, normative image of oneself as one would like to be. Death awareness, according to Jonas, brings life to focus as a moral project. While all forms of life exist on a scaled continuum of mediated world-relations, only humans are self-mediated. Self-mediation is the furthest expanse of freedom's horizon. Through this capacity for self-mediation, humans are able not only to imagine the external world of objects and to project that world symbolically, but also to imagine and symbolically project themselves. Humans therefore become multiple—subject and objects, concrete and abstract, seeing and seen, present and imagined.

The intensely world- and self-mediated character of human life, through which humanity gains causal, epistemic and moral leverage over the world constitutes for Jonas the uniquely open field of freedom through which the phenomenon of *moral responsibility* germinates. Though humans are part of the pageantry of life's forms, and though other forms of life alter the world, human capacities uniquely enable humans to imagine the future, to envision alternative courses of action and their various possible consequences, to intend and choose and evaluate, and thus to be morally responsible. In this sense, only humans can be morally in addition to causally responsible agents. Further, as Jonas writes, "The fact that [we] *can* assume [responsibility] means that [we] *are* liable to it."[1]

This connection between capacity and liability lies at the core of Jonas' thesis regarding the ontological grounding of ethics. The capacity for responsibility is ontologically grounded in human nature and is a correlate of the unique degree of human freedom. Through the intensely mediated character of human life, humans are capable of freely choosing courses of action, of being not only purposive but also intentional about which purposes can and should be pursued. And insofar as this degree of mediacy and the freedom that is a function of it are intrinsic to human nature, Jonas argues, as this chapter will proceed to explain, an ethics of responsibility is ontologically grounded in and morally beholden to the idea of humanity.

The foregoing paragraphs summarize my earlier chapters' account of the path Jonas charts from his philosophy of nature to his anthropology and ethics. The aim of this chapter is to provide a critical exegesis of the normative dimension of his ethics toward which this path has been leading, and to isolate its promise for the development of an ecotheological ethics of responsible participation. Toward this end, the next

1. Jonas, *Mortality and Morality*, 101.

section of this chapter will analyze Jonas's understanding of the significance of the future in his ethical vision, and then move to an exegesis of his imperative of responsibility and the norm through which it is expressed. I will then transition to an analysis of Jonas's conception of responsibility and the models of agency he advances as paradigms of responsibility. In the final section, I will provide a critical interpretation of Jonas's understanding of the role of theology in his work. Here I will argue that, against Jonas's own claims that theology is a "luxury of reason," theology plays a very constructive role in his project, even if it is not as integrated within his broader vision as it is in Gustafson's. I will indicate in my conclusion the ways in which his theological vision may be put to constructive complementary use in relation to Gustafson.

Futurity and the Critique of Previous Ethics

In order fully to appreciate the force of Jonas's ethics of responsibility, it is crucial to understand his view of the limited concern for the future in previous ethical thinking.

All previous ethical systems, according to Jonas, are inadequate to the contemporary ethos to the extent that they are calibrated to a scale of moral responsibility much too abbreviated for the powers that humans now possess.[2] As quoted previously, all previous ethical systems for Jonas are concerned with a "short-term context" deriving from the "short arm of human power." In other words, the limited consequential range of human action determines the relatively narrow domain of historical moral concern. The history of ethics is calibrated to a temporally contemporaneous and spatially proximate domain. But as his interpretation of contemporary technology concludes, the expanded efficacy of contemporary human power demands a compatible expansion of ethics' horizon of concern, especially with respect to the future.

---

2. Jonas's claim that there is no good precedent within the history of ethics for his vision of responsibility could be misunderstood to indicate that he does not think there is a sufficient precedent for responsibility in ethical thinking at all. But this is a significant misunderstanding. Jonas is fully aware that the idea of moral responsibility has a history. He is arguing, though, that there is little historical precedent for the contemporary scale of responsibility. The contemporary situation of human power demands ethical thinking to come to terms with the broadened range of the effects, and thus the stakes, of human action.

Jonas concedes that his claim that the history of ethics offers little in the way of contemporary guidance seems to be betrayed by foci operating in at least kinds of ethical traditions: the criterion of universalizability in Kantian ethics and the eschatological, future-oriented concerns of certain religious and political ethics. Jonas deems otherwise. With respect to Kant's ethics, Jonas argues that its limitation derives from, as he puts this, the strictly *rational constraint* imposed by the "can" operating in the imperative that one ought "Act so that you can will that the maxim of your action be made the principle of universal law."[3] If an action cannot be universalized, if the actor isolates the maxim of his action and therefore also him- or herself from a universal standard of moral evaluation, that action is morally wrong. The criterion of reason's consistency with itself in an action precludes rationalizing any act as a special case immune from moral critique. If a principle of action is to be judged morally right, according to Kant, it cannot be inconsistent with the maxims of action necessary to the ordering of a general moral community. While Kant's criterion of universalizability is perhaps a minimally necessary criterion for morals, what if the very possibility of moral action as such becomes threatened?

Jonas critiques Kant's imperative as inadequate for the present time insofar as it assumes the continued existence of rational human actors. The "can" appealed to in the imperative is that of reason's consistency with itself rather than the possibility of moral action as such. Jonas writes, "*Given* the existence of a community of human agents (acting rational beings), the action must be such that it can without self-contradiction be imagined as a general practice of that community."[4] For Jonas, this moral constraint is not sufficient for an age in which the future of humanity cannot be assured. The future existence of humans can no longer be assumed but is something that must be chosen. Commenting on this further, Jonas writes that, "The *presence of man in the world* had been a first and unquestionable given, from which all idea of obligation in human conduct started out. Now it has itself become an *object* of obligation: the obligation namely to ensure the very premise of all obligation . . . ."[5]

---

3. Jonas, *Imperative of Responsibility*, 10–11.
4. Ibid., 11 (italics original).
5. Ibid., 10 (italics original).

To say that the continued presence of humanity in the world is now an object of obligation is to acknowledge that human power has reached a point in which that presence cannot any longer be assumed. Weapons of mass destruction, nuclear, biological, and chemical are obvious examples of the capacity of technology to threaten this presence. Less obvious are the smaller more entrenched technological patterns in which human life is daily conducted. The threat of these entrenched patterns is not in their capacities for dramatic annihilation. Rather, the threat of "slow, long-term, cumulative change—the peaceful and constructive use of worldwide technological power . . . " is in the casual and seductive way in which it can eventually lead to "the overtaxing of nature, environmental and . . . human as well."[6]

Given these obvious and less obvious threats to the future of humanity in the world, Jonas deems that something like a preemptive imperative is necessary. His formulation of this adds substantive moral content to the formal character of the Kantian imperative. Positively, Jonas's imperative enjoins that we ought to "act so that the effects of action are compatible with the permanence of genuine human life." Negatively, it demands that we ought to "act so that the effects of action are not destructive of the future possibility of such life . . . ."[7] Jonas's defense of his imperative will be addressed further along. For now it remains further to distinguish his position from Kant's and then to examine his critique of selected future-oriented ethical systems.

As should be obvious by now, one of the distinctive marks of Jonas's imperative is that it no longer takes for granted the future presence of humanity in the world but seeks instead to preserve the possibility of that presence. In addition, the imperative adds normative substance to Kant's formal or procedural version. The future of humanity cannot be assumed, but ought to be preserved. An additional contrast concerns the relevant criterion of morality. The significant moral standard for Jonas is, in contrast to Kant's principle of universalizability, the consequential impact or effects of action. If the immediate or cumulative effects of an act undermine the possible future presence of genuine, or responsible, humanity, the act is morally wrong. It is, according to Jonas's imperative, morally forbidden to wager the idea of humanity with any act, even

6. Ibid., ix.

7. Ibid., 11. By "genuine" human life Jonas intends "responsible" human life, as suggested in ch. 3. This will be given more careful scrutiny further on in this chapter.

if the maxim of that act may be rationally universalizable. And thus Jonas's imperative replaces the Kantian formal criterion of universalizability, the internal rational consistency of a principle or maxim of an act, with the demand that action now cannot be incompatible with a human future. Jonas's imperative is in these ways fundamentally distinct from Kant's, though both his and Kant's imperatives reflect concern for the objectivity of morals.

With its orientation toward the future, Jonas's ethical vision also would seem to have something in common with other traditions in ethical thought, for example, with the ethics of certain religious and secular political eschatologies. It is important to consider Jonas's critiques of these traditions here since, in the concluding pages of this chapter, I will interpret how he draws in a qualified way on both a religious and a political vision to support his ethics. Of religious ethical visions toward the future, Jonas differentiates between moderate and extreme versions. Both versions seemingly qualify present action with a concern for the future. For Jonas, action in the present is viewed by eschatological religious ethics, in theory, not as leading to salvation so much as qualifying the agent for that future state. Indeed, action motivated by reward or the aim of salvation is typically taken morally and religiously to devalue the action. So the moral life of the religious agent is, ideally at least, not motivated by concern for reward or punishment. The specific prescriptions for the life pleasing to God vary of course according to the multiplicity of specific religious creeds. But the general virtues characteristic of the moderate religious life often do not significantly differ, according to Jonas, from classical non-eschatological ethics. Justice, charity, wisdom, and prudence, for example, are equally supportable by either religious eschatological systems or classical non-eschatological ones.

But in more extreme religious ethical styles, Jonas notes greater differences. In contrast to the more moderate religious style of life, the value of present action in the extreme religious life is more explicitly renounced. The extreme religious life, according to Jonas, is characterized by mortification and asceticism, by resistance to present desires and attachments. Jonas argues, however, that this denial of the present ironically betrays a concern for internal and this-worldly goal achievement. Conceived theoretically as a denial of the present, extreme versions of the religious ethical life actually indicate for Jonas an ethics of this-worldly self-perfection. The seemingly eschatologically funded ethical futurity of religious ethics, according to Jonas, actually reveals the priority of the present. Thus, Jonas claims not to find in religious ethics, at least not in his initial and admittedly superficial review of such ethics, the orientation toward the future that he deems necessary to a time of radical power. As I will explore later, however, Jonas does not dismiss the motivating powers of religious ethical visions. In fact, I will argue, one of the central features of his ethics is a speculative theological myth through which he embellishes the concern for the future demanded by contemporary moral responsibility.

Jonas also views the ethical visions of secular political eschatologies with suspicion. Whereas on his interpretation religious ethical systems betray, perhaps ironically, a concern for the present, he deems secular political eschatologies such as Marxism to be more truly future-oriented. Key to this claim is his understanding of the advent of the modern idea of progress. Jonas writes, "When [progress] . . . is wed with a secularized eschatology which assigns to the absolute . . . a finite place in time, and when to this is added a conception of a teleological dynamism which leads to the final state of affairs—then we have the conceptual prerequisites for a utopian politics."[8] The final good of the

8. Ibid., 16. It is important to acknowledge here the ways in which Jonas's critique of secular political utopianism relates to his critique of Gnosticism, which I interpreted in chapters one and three. There is a metaphysical dualism at work in both sensibilities that is morally problematic for Jonas, though it is perhaps more subtly manifest in its modern secular political incarnations. While the dualism operative in the gnostic vision leads to a radical axiological devaluation of the material, physical, and bodily and a privileging of the spiritual, political utopianism radically privileges a future state of affairs over the present. Extreme interpretations of both visions can yield not just devaluation but sacrifice of one side of the dualism for the other. While Jonas wants to preserve the emphasis on the future in political utopianism, his position is that, in light of the seductive but ambiguous power of humanity, the concern for the future needs to be normatively guided by a duty to ensure the ongoing conditions of responsibility. I will interpret this more carefully in the next section, focusing on his imperative. An additional similarity running through both political utopianism and Gnosticism concerns the privileged position of an elite who possess a special kind of saving knowledge, gnosis for the gnostics and for the political utopians right knowledge of the mechanics of political economy. The central difference of course is that the dualism in political utopianism, especially the Marxist version being treated here, is governed by a materialist secular metaphysics while gnostic dualism is religious. In addition to being concerned with the moral problem for political utopianism in a time seduced by the efficacy of technological power, Jonas strongly resists its dualistic metaphysics and is of course concerned to develop, as I have suggested elsewhere and as I will continue to explain, a more integrated metaphysics and ethics in which the exercise of power in the present is normatively guided by a concern for the vulnerability of the future. The

future state of affairs in such a view is historical rather than eternal, as with religious eschatologies. In such a conceptual schema, action in the present is undertaken explicitly as a means to this future good: "The obligations on the now issue from that goal . . . ."[9]

Though on Jonas's account the ethics of secular political eschatologies may contain the emphasis on the future that he deems crucial for the present time, the contemporary ethos also requires a radical suspicion of any utopian eschatology. In short, to the extent that Marxism's utopian goal has become a technological possibility, Jonas argues that it must be subject to strong critique: "A critique of utopia has become necessary with the seductive possibility of its realization. For the first time in the annals of man, thanks to the powers of technology, the dream appears to be capable of turning into a task . . . ."[10] "Nothing," writes Jonas, "could tempt Prometheus unbound more than the dream of the highest earthly good believed within its reach, and nothing can become more dangerous to mankind than a mistaken pursuit of it."[11] The future orientation of secular political eschatologies is highly dangerous insofar as it is guided by an overly optimistic sense of the efficacy of power to achieve the future final good. Though a truly responsible ethics demands that present action be normatively guided by concern for the future, it also requires that the means toward that future be checked by caution rather than zealotry, or in Jonas's terms, by a "heuristics of fear" rather than hope. The necessity of this constraint is demanded by the moral ambiguity of contemporary human power.

In a time of radical power, the future should not be rushed into optimistically. A norm is needed to constrain the impact of present action on the future, especially in a time in which the seductive efficacy of technology is so morally ambiguous. The force of constraint required by an ethics of futurity in a technological age, for Jonas, must equal the enormous force and ambiguity of contemporary human power. In a startling move, but one that has been implied in my previous chapters, Jonas turns for this principle of constraint to the vulnerability and per-

---

connection here between Gnosticism and political utopianism will be addressed again in my conclusion where I will offer a critical interpretation of Jonas's own political vision as in some ways capitulating to the utopian political problematic he is here so concerned to avoid.

9. Jonas, *Imperative of Responsibility*, 17.
10. Ibid., 178.
11. Ibid.

ishability that are the hallmarks of life's past, present, and future. The recognition of life's intrinsic fragility, its intransigent precariousness, and a mindset of suspicion or fear rather than optimism are prerequisite to Jonas's ethics. Jonas's philosophy of organism is foundational to this mindset, rooted as it is in life's fundamental metabolic striving, its needful freedom. Understanding the connection Jonas forges here between life's vulnerability and his cautionary ethics requires moving into a deeper analysis of his imperative of responsibility.

Jonas's Imperative of Responsibility

In its positive formulation, Jonas's imperative states that we ought always to "[a]ct so that the effects of action are compatible with the permanence of genuine human life."[12] There is of course much to consider in an exegesis of this norm. At the outset, it is important to note that its concern with the consequences of action is governed by a duty to the future of genuine human life. I will engage this in more detail further along in this section, but Jonas affirms his imperative as categorical rather than hypothetical. It is neither purely consequentialist, nor purely deontological. It stresses that the consequences of acts ought to be "compatible" with the possibility of a genuinely human future, that duty to such a future provides the normative constraint on action. Just which acts may or may not be compatible with such a future depends on understanding what Jonas intends by "genuine" human life. For Jonas, the peculiarly human form, or "genuine" human being, is to be responsible. And so to act in a way that is compatible with a genuinely human future demands acting in ways that are compatible with the conditions of responsibility, which entails acting in ways that do not degrade the natural conditions through which the capacity for responsibility has emerged. Thus, as I have expressed this elsewhere, while Jonas's imperative is anthropocentric, it is a qualified anthropocentrism.

In addition to these points, Jonas's imperative makes it a duty for agents to imagine and to feel possible futures. Because the consequences of technology cannot be precisely calculated, the work of imagination becomes an obligation. In dealing with the colossus of technology as Jonas conceives it, the aim is to constrain possibly devastating outcomes before they become actual. Recognizing and obeying the moral claim of the future of humanity in the present requires the labors of imagination

12. Ibid., 11.

and reason to envision and affectively to relate to the possible scenarios of technological action that may threaten that future. This anticipatory duty requires agents to deploy the fullest powers of their cognitive, imaginative, and affective capacities. The duty to anticipate is a primary duty within the imperative, for the strength of the call of the future requires fully conceiving the vulnerability of the future. I will indicate in the final sections of this chapter, and argue more fully in my concluding chapter, that Jonas most explicitly illustrates the significance of imagination in his ethics in his theological writings, and that this is one of the important reasons for questioning his claim that theology is peripheral to the defense of his ethical vision.

The normative content of Jonas's imperative needs to be more fully interpreted, to say nothing yet even of its justificatory warrant or possible critiques of it. The content of Jonas's norm can best be understood by turning to the way that Jonas responds to the question of why the future of human life ought to be preserved. Thus far it has been shown that he perceives the future of life to be threatened by the enormous powers of humanity. But, given this, why ought humanity to continue to exist? It seems that the planet would be much better off without humans. Is not the imperative to preserve the future of humanity overly anthropocentric, simply extending the career of a species that poses the gravest threat to the planet? Why ought not "the existence or essence of man as a whole be made a stake in the hazards of action"?[13]

Answering these questions, according to Jonas, redounds ultimately to the primordial metaphysical question of why something is rather than nothing. For Jonas, the logic of this question is not causal, but moral. He is not concerned here with efficient but with final cause. In other words, any efficient causal answer to why something is only regressively begs the question of the cause of the cause. For Jonas, the real burden of the question is why there ought to be anything rather than nothing. Given this understanding of the question and the importance of it to a justification of his imperative, Jonas's project depends on nothing less than a whole metaphysics of nature, centering on the philosophy of biotic purposiveness interpreted in chapter three. Thus, Jonas's ethical imperative is metaphysically contingent. Ironically, according to Jonas, the preservation of all of life, each form of which is an end unto itself, depends upon preserving the future of genuine humanity from

13. Ibid., 37.

the threat of present humanity. If Jonas's ethics is anthropocentric at first glance, it is instrumentally so. Humanity must be preserved, ultimately, for the sake of all life's future.[14]

Jonas highlights two mutually reinforcing aspects of his imperative. First, there is the duty to ensure a future for genuine humanity. Second, there is a duty to ensure the conditions of life necessary to this idea. Though these are intrinsically related, Jonas grants priority to the first dimension rather than the second. In other words, one should not excuse oneself from the duty to future humanity given a judgment about the possibly unpropitious future quality of human life on earth. One should not decide, given the present and likely future state of the world, that it is better not to bring children into the world. But what is the reasoning behind this principal duty to posterity? Jonas argues that the priority of the duty to ensure a genuine human future derives from the demand imposed by the radical threat of human powers today. An objective norm must be consulted that transcends our wishes and the wishes of future humans. What is to be referenced as normative now is an "ought" arising from within the idea of genuine humanity that transcends present human wishes and the anticipated desires and sentiments of future humans. As Jonas puts this, the priority of the duty to posterity "means that in the final analysis we consult not our successors' *wishes* (which can be of our own making) but rather the 'ought' that stands above both of us. To make it impossible for them to be what

14. Richard Bernstein, one of Jonas's former colleagues at the New School for Social Research, has critiqued Jonas's ethics as having an "unresolved tension" at its core, a tension between his critique of the Western philosophical tradition's anthropocentrism and the rather anthropocentric formulation of his own imperative of responsibility. According to Bernstein, this is reflected as a "gap" in Jonas's thinking between the "imperative to preserve organic life" and the "more specific conclusion that the primary object is to preserve human existence." For Bernstein, "if one is to move from the premise that there is a supreme obligation to preserve the conditions for the possibility of organic life to the conclusion that 'the existence of mankind comes first,' then one needs an independent argument to justify this strong conclusion. It is this argument that I find lacking in Jonas . . . ." See Bernstein, "Rethinking Responsibility," 13–20. But as I interpret Jonas, and I think he is himself very clear about this, the argument for the priority of human existence turns on a respect for the innate goodness of the natural processes through which the human species has emerged. Thus the imperative to preserve genuine human life does not evince a gap in Jonas's argument so much as the deep roots of his norm in his broader philosophy of nature. For this reason, I think it is accurate to describe Jonas's ethics as an "instrumentalized," or, as I elsewhere put this, as a "qualified" anthropocentrism.

they *ought* to be is the true crime, behind which all frustration of their desires, culpable as it may be, takes second place."[15]

This duty to make it possible for future humans to be what they ought to be, a profound responsibility indeed, of course rests on a commitment to a particularly normative idea about what genuine humanity means. The complexity of this idea and the deftness of Jonas's articulation of it, which I began to interpret in the chapter on his fundamental ideas, warrant quoting him at length. Jonas writes that the duty to the future idea of genuine humanity means,

> that it is less the *right* of future men (namely, their right to happiness, which given the uncertain concept of "happiness," would be a precarious criterion anyway) than their *duty* over which we have to watch, namely, their duty to be truly human: thus over their *capacity* for this duty—the capacity to even attribute it to themselves at all—which *we* could possibly rob them of with the alchemy of our 'utopian' technology. To stand guard over this onerous endowment of theirs is *our* cardinal duty toward the future of humanity as such.[16]

Thus, basic to the idea of humanity according to Jonas is the capacity to recognize and to respond to duty. Enfolded in the duty to ensure a genuine human posterity, then, is the duty to ensure the continued existence of the capacity for duty that is constitutive of what it means to be human.[17] Again, in Jonas's words, the present duty of humanity is to

---

15. Bernstein, "Rethinking Responsibility," 41–42 (italics original).

16. Ibid., 42 (italics original).

17. Richard Bernstein raises another critique of Jonas that my exegesis of Jonas on this point clarifies. Bernstein writes that even if Jonas has supported his move from the duty to preserve life to the priority of human life, which I have argued that he has, "there are many ways of preserving the continuation of life—even human life—on this planet which we would find morally repulsive, even, morally irresponsible. If for example, we recognize, as Jonas does, that one of the most serious threats to the continuation of future life is overpopulation, then we might tolerate or encourage famines and the spread of certain diseases that can so effectively reduce populations." Bernstein acknowledges that there is plenty of evidence that Jonas would resist such measures, but suggests that as Jonas's imperative stands, in order to support such resistance his ethics of responsibility "demands the acknowledgment of other supplemental imperatives and principles." See Bernstein, "Rethinking Responsibility." I do not take issue with the point that, in the face of the many moral challenges known and unknown in a technological time, Jonas's ethics could be strengthened by supplemental imperatives. But I think that as it stands Jonas's imperative makes the preservation not just of life, nor just of human life, but *responsible* human life a priority, and in so doing contains

"tax" future humans with "the very sort of existence that is capable of the burden which is the true object of our duty to bestow that existence .... That there *be* future bearers [of this burden], so as to ensure [this burden's] very perpetuity in the world, comes first in the hierarchy of duties."[18] But the priority of this duty leads to a secondary duty, through which the seeming anthropocentrism of Jonas's ethics becomes radically qualified.

The duty to ensure the perpetuity of the burden of duty "presupposes that we have not prejudiced [the future's] capacity to bear it." Not undermining the capacity to bear the burden of duty, to recognize and respond to duty, is the normative constraint that is to guide actions in the present that effect the future conditions of human existence. Thus, the second aspect of Jonas's imperative makes it a duty to preserve the right qualitative conditions of genuine human existence. The primary duty to ensure humanity's continued existence determines the secondary duty to preserve the conditions necessary to the future idea of humanity. In other words, the second obligation is to ensure conditions of existence that allow future humans to be what they ought to be. The duty to ensure the posterity of the future is not a duty to protect the right to exist of future individual humans or to secure their wishes for happiness. Rather, it is to ensure the continued capacity for duty, or the idea of humanity. Jonas stresses that his point is not that *if* there are future humans then their conditions of existence ought to be such and such. His imperative is categorical, and his point is that *because* there ought to be future humans, the human duty now is to ensure the perpetuity of the essence of the idea of humanity, or the conditions of the capacity for duty.

Obviously, Jonas here aligns himself in some way with the Kantian distinction between hypothetical and categorical imperatives. But there is, as already mentioned, a key point of difference. Of the difference in his imperative, Jonas writes,

---

within itself a basis upon which to argue against planned famines and epidemics to control overpopulation. Such measures would at the very least corrode the conditions for a future of *responsible* humanity insofar as they would compromise, for example, the conditions of trust, care, and mutual concern upon which, among other conditions, Jonas's understanding of responsibility depends.

18. Ibid. (italics original).

[s]ince *its* principle is not, as with Kant, the self-consistency of reason giving to itself laws of conduct, that is, not an idea of *action*...but is rather the idea of possible agents in general, for whom it claims that such *ought to exist*...and is thus 'ontological,' that is, an idea of being—it follows that the principle of an 'an ethic of futurity' does not itself lie *within* ethics as a doctrine of action . . . but within *metaphysics* as a doctrine of being . . . .[19]

In contradistinction to Kant, then, Jonas grounds his categorical imperative in the ontology of the human agent as that ontology is influenced by his metaphysics of nature. Practical reason for Jonas, then, is not primary as it is for Kant. Practical reason and the objective duty guiding it, instead, depend upon the metaphysical ventures of theoretical reason.

As mentioned earlier, and this is more obvious now, Jonas's ethics works against the grain of dominant tendencies in recent philosophical ethics. Specifically, Jonas's views challenge the modern denial of metaphysics and the dogma that "is" and "ought" cannot be bridged. Jonas flatly rejects these dogmas head-on by claiming metaphysical support for an ethics in which "ought" derives from "is." That humanity *ought* to be preserved, according to Jonas, derives, ultimately, from the fact that nature *is* purposive and has brought about purposive humanity. The "*is*" of being includes an "*ought-to-be.*" The following section will interpret Jonas's ontological grounding of his imperative through his negotiation of these two modern dogmas. This will, in part, provide the groundwork for my analysis of his account of the meanings and paradigms of responsibility, and also point forward to my concluding argument that his theology plays a crucial role with respect to his claim, to be discussed in what follows, that being's goodness is intuitively evident.

Ontological Grounding of Responsibility

Jonas's case for the ontological grounding of value in the nature of things is pivotal to his ethics. The non-derivability of "ought" from "is," he claims, which would seem to pose a significant obstacle to his argument, is warranted only by a presupposed neutralization of being. The asserted logical chasm between "is" and "ought," he argues, is itself grounded in a particular metaphysics. The operative metaphysics in this case is the reigning world-picture underlying positivist science in which being is theorized as value-free and non-purposive. That this is the case

19. Ibid., 43–44.

ironically implies that adherence to the dogma of non-derivability conflicts with the dogmatic modern denial of metaphysics. According to Jonas, then, these dogmas are mutually contradictory. Problematic beyond this contradiction is the fact that the denial of metaphysics is tethered to and circularly determined by a specific epistemology. "Just as the dogma of 'is and ought' presupposes a definite concept of being," Jonas writes, "so does the denial of metaphysical truth presuppose a definite concept of knowledge, for which it is indeed true: 'scientific' truth is not to be had about metaphysical objects—[and this is] once again a tautological conclusion since science is just concerned with physical objects."[20] It follows that the possibility of metaphysics remains open if the positivist epistemology does not fully account for all that can be known and how.

Jonas's ethics is self-consciously and explicitly rooted in a metaphysics of purposive nature. He writes, "for the sake of our first principle (which should tell us why future men matter by showing that 'Man' matters), we cannot avoid taking the imprudent plunge into ontology . . . ."[21] Why is this plunge imprudent? Because Jonas acknowledges that it may be the case that the ground he seeks for his principle may remain shrouded in an "abyss of the unknowable." It is for this reason that he suggests religious faith has a certain strategic advantage over the work of philosophical reason. And yet, Jonas claims, signaling his views on the role of theology to be taken up later in this chapter, though "faith in revealed truth [may] very well supply the foundation for ethics . . . it is not there on command, and not even the strongest argument of need permits resorting to a faith that is absent or discredited."[22] Given this claim, reason must take the plunge into ontology in the burden of its quest for an adequate foundation for ethics. Despite Kant's claim to the contrary, Jonas argues, "for the

20. Ibid., 44.
21. Ibid., 45.
22. Ibid. While Jonas claims this, my argument is that he himself, in the end, "re-sorts" to theological speculation in support of his ethics. I will suggest that this may not merely be a "strategic" move in response to "need", but one that he comes to see, for reasons I will explain further along, as necessary. On this I differ from Lawrence Vogel's judgment, more in keeping with Jonas's own claims, that while a theological vision and a stance of religious faith may "complement" Jonas's ethics the ethics does not depend on theological commitments. See Vogel, "Does Environmental Ethics Need a Metaphysical Grounding?"

search to be unprejudiced, the worldly philosopher struggling for an ethics must first of all hypothetically allow the *possibility* of a rational metaphysics . . . if the rational is not preemptively determined by the standards of positive science."[23] Jonas's "worldly, unprejudiced" quest for the ontologically grounded value of humanity as such, supported by a metaphysics of purposive nature, as suggested earlier, entails angling for a response to that most basic of metaphysical questions, "Why ought there to be anything rather than nothing?" Either being ought to be or not, and if it ought to be, this must be because being as such is good. In short, at its ultimate level according to Jonas, the question of whether humanity ought to be turns on the primordial question about the ontological status of the good.

The two questions concerning why humanity or any particular entity ought to exist and why being in general ought to be are related but distinct, and for Jonas the question regarding being in general is primary. The answer to the question of why any particular entity ought to exist, according to Jonas, can only be relative, depending as it does on a comparison between degrees of actual alternatives possible in reality as it is given. If some particular thing does exist, then its existence is better than the alternative and therefore it ought to be. And, further, that some particular thing does exist and thus ought to exist allows for discrimination in terms of degrees of quality of existence. But the question of why there ought to be anything at all, rather than nothing, is much more radical and can only be answered absolutely. This is so for the fact that the question of why being in general rather than nothing ought to be, cannot be answered in terms of degree. There is only being or nonbeing. But a sufficient answer to this question cannot be taken on faith, argues Jonas. On Jonas's view, referencing, for example, the monotheistic doctrine of creation regressively begs the question in the way discussed earlier. While the doctrine of creation answers the question of why being is by referencing divine causation, the question of God's existence and why God ought to exist remain open. According to Jonas, then, "the question of whether the world ought to be . . . can be separated from any thesis concerning its authorship."[24]

The question of why there ought to be anything rather than nothing, for Jonas, is essentially a question for reason, regardless of how one

23. Jonas, *Imperative of Responsibility*, 45.
24. Ibid., 47–48.

views the status or validity of religious knowledge and faith. Answering this question is a task that, according to Jonas, has "always belonged to metaphysics and to metaphysics alone—under the condition of belief no less than of unbelief . . . ."[25] And again, the question is not asking about efficient cause: "The meaning of the question why there *is* something and not nothing must be this: Why *ought* there to be something in preference to nothingness, whatever the *cause* of its coming to be might be?"[26] The significance of the "ought" here is that it focuses attention on the question of "whether there *is* such a thing as 'value,' not as something here and there actual, but as something possible in its very *concept*."[27] In short, as I suggested above, this whole train of reasoning means that if providing a justification for the imperative that humanity ought to exist entails passing through the question of why being ought to be over nonbeing, then this justification redounds to whether or not the good is objective.

Pursuit of the objectivity of value is not only a logical but also, and most importantly for Jonas, a moral demand. That there are subjective values at work in the world is incontrovertible, but these are not sufficient to the moral task of a technological age as Jonas envisions it. Neither the needs, nor desires, nor the preferences of human individuals or communities can provide the normative constraint of power that contemporary technological efficacy requires. For, "Only from the objectivity of value could an objective 'ought-to-be' in itself be derived, and hence for us a binding *obligation* to the guarding of being, that is, a responsibility toward it."[28] But Jonas's claim that an adequate imperative demands that this route be taken, and that the gravity of present moral situation demands a strength of obligation that can only be secured through an objective foundation, is not the same as demonstrating the link between value and being. That something is needed does not make it so. But Jonas has, by way of a methodology that allows him to posit the subjective value of individual organisms, provided a way, if not to demonstrate, at least to suggest the plausibility of the conviction that the good is objectively grounded.

25. Ibid., 48.
26. Ibid.
27. Ibid., 49.
28. Ibid., 50 (italics original).

As discussed in chapters one and three, Jonas's existential interpretation of biological facts leads him to claim that purposiveness exists rudimentarily in life's most primitive forms. Purposiveness is not unique to humans. All forms of life are goal-directed if only in the sense that metabolically all organisms physiologically affirm their being in relation to nonbeing. Purposiveness is present wherever and whenever particular organisms exist. Jonas writes, "Nature [in its various specific organic forms], by entertaining ends . . . also posits values. For with any *de facto* pursued end . . . attainment of it becomes a good, and frustration of it, an evil; and with this distinction the attributability of value begins."[29] Value is present and attributable wherever there is purposive pursuit of some end. But this account of value only attests to its presence internally within the dynamic of an individual organic subject and its specific purposes.

Left unanswered is the question of the objective goodness of purposiveness as such. Purposiveness, Jonas has argued, is engrained in the essential striving, the needful freedom of particular organic entities. But is the good engrained in the nature of things? Whether or not the good is ontologically grounded depends on whether purposiveness in general is good. And until the goodness of purposiveness as such is granted, no objectively grounded "ought" can be claimed to derive from being.

Jonas does not rationally demonstrate the move from the value relative to purposive organic individuals to the absolute, objective good of purposiveness itself. In contrast to the dynamic of purpose and value in individual organisms, which Jonas argues is empirically supportable and logically necessary, the question of the goodness of purposiveness as such cannot be rationally proved. Reason must give way to intuition. He writes of this that, "We can regard the mere *capacity* to *have* any purpose at all as a good-in-itself, of which we grasp with intuitive certainty that it is infinitely superior to any purposelessness of being . . . there is plainly no going back behind [this intuition] for something more basic to underpin it."[30] For Jonas, then, intuition grasps the axiom that the very capacity to have purposes is good, and thus that the good is ontologically grounded. Though the goodness of purposiveness itself

---

29. Ibid.
30. Ibid., 80.

"is a matter of ultimate metaphysical choice which can give no further account of itself . . . still it commands an evidential intuition of its own . . . ."[31] This "evidential intuition" is that being in general provides its own testimony to its goodness insofar as every individual purposive being instantiates its value by affirming itself against nonbeing. Through the particular affirmations of beings for themselves, being as such can be intuited to be declaring itself for itself. In defense of this intuition, Jonas suggests that there is no logically self-refuting alternative to affirming being's declaration of its goodness. He writes, "Against this verdict of being there is no counter-verdict, for even saying 'no' to being betrays an interest and a purpose. Hence, the mere fact that being is not indifferent toward itself makes its difference from nonbeing the basic value of all values . . . ."[32] Goodness is immanent to being insofar as being in its essence cannot be indifferent to itself, but purposively affirms its own goodness through every subjectively valued purpose.

Once consciously registered by the will of humans who can recognize their radical impact on the nature and course of being, and thus can intentionally aid or impede being's own project, being's primitive expression of its goodness issues a claim to moral consideration. As Jonas puts this, "whenever this first, self-validating good happens, in any of its individuations, to come under the custody of a will, it addresses an 'ought' to this will."[33] Being's own willing of itself against nonbeing, and thus its expression of its goodness, is translated into an "obligating force in the seeing freedom of man, who as the supreme outcome of nature's purposive labor is no longer its automatic executor, but with the power obtained from knowledge, can become its destroyer as well."[34] Aside from the presence of humanity, in whom nature's purposiveness has become most intense, whose freedom has reached a degree of mediacy unknown to other forms of life, being's own goodness is, so to speak, self-preserved. Being takes care of itself.

Preserving being's goodness does not become a moral task except in the context of human freedom. Articulating the transition to conscious moral obligation from being's unconscious willing of itself and

31. Ibid.
32. Ibid., 81.
33. Ibid., 80.
34. Ibid., 82.

thus of the security of its goodness constitutes the foundational problem of a contemporary ethics according to Jonas. In response to this foundational problem, Jonas has pushed the limits of his naturalistic philosophy, arguing in the end that human moral obligation to being as such rests on intuition. Before moving to an interpretation of Jonas's theological writings, which I take to be at least in part motivated by his dissatisfaction with his argument here, it is necessary, for the purpose of specifying both the insights and liabilities of his imperative of responsibility, to turn to a close analysis of his account of responsibility's structure and ideal paradigms.

Responsibility: Analytic, Models, Imperative

After having interpreted the formal and normative dimensions of Jonas's imperative and after having examined his argument for the ontological grounding of obligation, it remains to analyze his understanding of the structure of responsibility and to interpret his view of its exemplary paradigms. In elucidating the concept of responsibility, Jonas notes three conditions of responsibility and two basic senses of it. The first and most basic condition of responsibility concerns causal *efficacy*. Responsibility arises as a result of having the causal power to alter the world in some way. The second condition concerns *control*. To be responsible for causal efficacy entails that the actor have some control over his actions. The third condition concerns *knowledge*. The actor must be able to foresee in some measure the consequences of the power under his command. In sum, the three basic conditions of responsibility are efficacy, control, and knowledge. Given these conditions, there are possible two different senses of responsibility, formal and substantive. Formally, there is the sense of responsibility as accountability *for one's acts*. Substantively, there is responsibility *for objects* that, by the nature of their relation to agents, possess a claim to moral consideration. Together, these conditions and senses constitute the gestalt of Jonas's concept responsibility. In what follows, I will examine more closely the crucial distinctions between the formal and substantive aspects.

Formally, Jonas claims, accountability for deeds is more of a legal than a moral notion. In this formal-legal sense, the agent is answerable for his acts. Liability is attributable to the agent insofar as there is a demonstrable and direct causal connection between his deeds and their effects. This sense of responsibility is not yet moral, though it suggests a precondition of morality. The emphasis at this point is on the possibility of directly linking outcomes to the causal efficacy of some agent. In other words, the formal sense of responsibility depends principally on responsibility's first condition of possibility, efficacy. Jonas writes, "So understood, 'responsibility' does not itself set ends or disallow ends but is the mere formal burden on all causal acting among men, namely, that they can be called to account for it."[35] The transition from this formal sense of responsibility to its substantive moral valence depends on analysis of the agent's volitional and cognitive relation to the act. The question of moral responsibility requires that the formal criterion that effects can be attributed to an actor, but this is not a sufficient, only a necessary precondition for moral responsibility. Moral responsibility "concerns not the *ex post facto* account for what has been done, but the forward determination of what is to be done . . . ."[36] In other words, the substantive moral sense of responsibility includes not only being able to link effects to the cause of an agent, but also and more importantly an analysis of the relation between that efficacy and the agent's control (the volitional relation) and knowledge (the cognitive relation). Thus the second and third conditions of responsibility are emphasized in substantive moral responsibility.

Substantive moral responsibility emphasizes not "what has been done" but "what is to be done." By this "forward determination of what is to be done," the agent is responsible "not in the first place for [his] conduct and its consequences but for the *matter* that has a claim on [his] acting."[37] The *matter for which* an agent is morally responsible is external to the agent yet within the range of his efficacy. An agent is responsible for *what is to be done* insofar as his doing can impact some thing external to him that has a claim on him and is in some way dependent on his efficacy. The dependent moral object commands consideration and enjoins a two-layered commitment of the agent, objective and subjective. First, the agent is "objectively responsible for what is . . . entrusted [to him]," and second, the agent is "affectively engaged through the feeling

---

35. Ibid., 92.
36. Ibid.
37. Ibid. (italics original).

that sides with [that objective responsibility]."[38] Jonas puts this well in claiming that the objective and subjective layers of responsibility coincide in the "right-plus-need of the object . . . ."[39] The objective aspect of moral responsibility stems from the efficacy of the agent *to impact* an object that has a claim to be; the subjective from the agent's *care* for the object with a claim: "First comes the 'ought-to-be' of the object, second the ought-to-do of the subject who, in virtue of his powers, is called to its care."[40] In sum, the ethics of responsibility Jonas aims to articulate, the elucidation of responsibility that he deems to be demanded by the unique pressures of a time of radical power, entails forging a tight connection between the formal-objective and the substantive-subjective

38. Ibid.

39. Ibid.

40. Ibid., 93. It is important here briefly to note Jonas's understanding and use of the term "care" in contrast to Heidegger's. Without moving into anything approximating an adequate discussion of Heidegger's concept of "care" (*Sorge*), a very rough excursus is warranted. For Heidegger, *Sorge* refers to the fundamental mode of Dasein's being-in the world, designated by the difficult formulation: *being-ahead-of-itself/being-already-in-(the world)/being-alongside-entities (and caring-for-others)*. See Heidegger, *Being and Time*, 237. As I understand this, "being-ahead-of-itself" refers to Dasein's comportment toward its ownmost possibilities, it is being free for and toward the potentiality-for-Being; "being-already-in" refers to the situation of Dasein in the world, a situation already there into which Dasein is thrown and which precedes and grounds Dasein's openness to possibilities; "being-alongside-entities" to the inclusive field in which Dasein works out its "being-ahead-of-itself" and "being-already-in."

Now, for Jonas, the notion of "care" is deployed in a way that draws attention not to a description of the way human beings are but to a prescription for the way humans ought to be. As will be discussed further along, the way humans ought to be is exemplified by the parental and political paradigms of responsibility: what is prescribed is a total, continuous, future-oriented, non-reciprocal relation of care, not to abstract other "entities," but to the vulnerability of those concrete others, whether child or citizen, that one has come into a custodial relation with, whether by nature or convention. Thus in contrast to the horizontality of Heidegger's "being-alongside," at the core of Jonas's understanding is a vertical or hierarchical arrangement between the one who cares and the one's cared for. The linkage between Jonas's conception of care and his philosophy of nature reveals an even more profound contrast. This is expressed in Jonas's claim that, "if the deeper insight of Heidegger is right—that, facing our finitude, we find that we care, not only whether we exist but how we exist—then the mere fact of there being such a supreme care, anywhere within the world, must also qualify the totality which harbors that fact, and even more so if 'it' [nature] alone was the productive cause of that fact, by letting its subject physically arise in its midst" (*Phenomenon of Life*, 234). With this claim, it is again evident that while Jonas is indebted to Heidegger in many ways, his project is at the same time a significant departure from him: Jonas attempts to provide an ontological grounding for ethics rather than a fundamental ontology.

relation between the causal powers of agents and the sentiments that bind agents to vulnerable objects with moral claims.

Given this analytic of responsibility, with its requisite *conditions* and its possible objective and subjective *valences*, Jonas articulates a further crucial distinction between *modes* of responsibility. His unfolding of these modes provides the basis for his account of two fundamental paradigms of responsibility, the parent and the politician. These different modes arise either through circumstance or contract, by way of natural order or convention. In either mode, the agent's moral responsibility includes both objective and subjective aspects. That is, the objective dimension of the agent's responsibility stems from the fact of his power *over* the object, the subjective dimension from his felt obligation *to* or *for* it. But this power over and obligation for can arise in different ways and result in different modal embodiments of responsibility. One fundamental commonality between these modes, according to Jonas, is that the moral relation between the agent and the object is non-reciprocal. Objective responsibility, as just interpreted, derives from recognition of the object's dependence or vulnerability in relation to the power of the agent, which power leads to subjective-felt responsibility for it. Jonas marks the contrast between the ethics of reciprocity and non-reciprocity by referencing the one domain as the concern of horizontal relations, for example between contemporary equals, and the other as vertical relations, for example between parent and child. Reciprocal and horizontal ethical relations are, according to Jonas, conditional and occasional in contrast to the unconditional and continuous ethical claims that absolutely and vertically bind agents to certain objects. At issue in this difference is the nature of the moral object and the way that this nature structures a specific form of relationship to an agent. It is important to note that Jonas in no way dismisses the importance of ethical reciprocity and the types of relationships such an ethics aims to preserve. He is preparing the groundwork for his ethical theory's focus on the future—a focus on the non-reciprocal claims generated by future moral objects dependent upon efficacious agents in the present.

Now, within the class of non-reciprocal moral relations with which Jonas is concerned there are two distinctive modes of responsibility. These modes are differentiated in terms of how the non-reciprocal relation of responsibility arises. One mode is natural, the other conventional. The mode of responsibility is determined *a priori* in the cases of natural relation between agents and dependent objects, whereas it is determined *a posteriori* in conventional cases. Jonas states the distinction in the following way: in the case of natural responsibility "the immanent 'ought-to-be' of the object claims its agent *a priori* and quite unilaterally," and in the case of conventional responsibility the claim of the object is *a posteriori* and turns "upon the fact and the terms of the relationship actually entered into."[41] The claim of the object on the agent in the natural case is given; the claim in the conventional case is the result of contract in which the moral object assents to some other's power. Jonas illustrates these modes of responsibility by turning to the paradigms of the parent and the statesman. These modes are distinct and yet share features in common that propel Jonas's further elucidation of his imperative of responsibility.

With respect to their differences, the natural responsibility that arises in the relation between parent and child is given, whereas the responsibility of the statesman for a populace is contracted. The dependent objects of the one mode are few and intimately known, of the other many are unknown; in the natural case the object is present in flesh, in the conventional in abstract idea. In short, the "most basic naturalness of the one contrasts with the utter artificiality of the other . . . ."[42] Yet in spite of these differences, argues Jonas, both modes are characterized by obligations to the "total," "continuous," and "future" welfare of other human beings.

By the "totality" of parental and political obligation, Jonas refers to both the objective and subjective aspects of these paradigms of moral responsibility. Objectively, the parental and political responsibility concern not just momentary or limited concerns with their objects, but instead encompass the whole being of their objects. With respect to parental responsibility, the object of concern is not merely the child's immediate needs or desires but the whole of the child's physical and mental being, and ultimately a concern for the complex flourishing of the child. Drawing on Aristotle's understanding of the purpose of the state, that the state is to preserve the possibility and the good of the common life, the statesman's responsibility is similarly total in nature. Subjectively, the responsibility of the parent and the statesman share a

---

41. Ibid., 95.
42. Ibid., 98.

common sentiment in relation to their objects. The vulnerability and precariousness of the child and the state generate a robust and commanding sentiment of care. While there are of course important differences between the models of the parent and the statesman, as mentioned above, Jonas argues that both the parent's and the statesman's responsibility is objectively and subjectively total.

Continuity and futurity of responsibility follow from the total nature of parental and political responsibility. In neither case is responsibility limited to a definite time or space. "Neither parental nor governmental care can allow itself a vacation or pause," Jonas writes, "for the life of the object continues without intermission, making its demands anew, time after time."[43] This continuity distinguishes these models of responsibility from the nature of responsibility characterizing other agential roles. The responsibility of the day care provider or the platoon leader, for example, may similarly be total but is limited to specific frames of time. But, while the case for the totality of parental and political care seems reasonable, is this so for continuity and futurity? For the child eventually becomes adult, growing beyond the dependence on his parents that initially characterizes his relation to them. And political communities can endure beyond the individual lives of their leaders. Further, in both cases, the future is neither fully knowable nor under the control of the parent or statesman. So the reality of a break in continuity and the indefiniteness of the category of the future seem to pose a problem. How can one be responsible for what has and ought to outgrow dependence, or for what one should not determine and therefore for what one cannot know?

Jonas's response to these possible critiques is that instead of weakening his argument they strengthen it. It is true both that the relationship between parent and child and state and statesman will change and that the futures of the child and the state cannot be fully predicted. And these changing relationships and the unknowability of the future are in part the intended outcome of responsible parenting and statesmanship. After all, if a parent has parented well and if a statesman has led well, the child and the state will have gained their own capacities for responsible freedom. Precisely this, the developed capacity for responsible freedom, argues Jonas, defines the horizons of continuity and futurity

---

43. Ibid., 105.

of parental and political responsibility. It is precisely the enabling of this "transcendent horizon" through which the child becomes adult becomes citizen—a development in which the objects of parental and political responsibility interpenetrate—that the continuity of care and the concern for the future must respect.

It is important now to draw together the significance of Jonas's analytic of responsibility and his models as they are enfolded within his imperative, for the purpose of better understanding his ethics of responsibility toward the future idea of humanity. The objects of moral concern for the parent and statesman share in the vulnerable, transient nature of all forms of life. But in addition, they share with the agent his humanity. Earlier, Jonas defended the idea that all organisms are ends in themselves. Every particular organism has its own end—to be rather than not to be—expressed through its needful freedom. While all organisms hold in common the end to be, the purposive path toward the attainment of that end cannot but be species specific. Humans cannot share in the ends of other nonhuman organisms. But, uniquely within the panoply of life, humans can be responsible for them. The purposes of nonhuman organisms are not the same as human ends, though humans can "guard their self-purposes."[44] In the interhuman spheres of the parent and statesman, on the other hand, the agent can include the end of the object within his own end. This shared humanity, and the possibility of including another's end within one's own, means that responsibility on the interhuman level can be reciprocal. In the cases of both the parent and the statesman, then, the responsibility relation contains within it the possibility of reversal. This cannot be the case between present and future humans and between humans and other forms of life, since neither future humans nor other forms of life can reciprocate responsibility. Of course, this does not mean that other forms of life, or the whole of life itself, cannot be moral objects to which human agents are responsibly obligated. Instead, as the following will show, it means for Jonas that the force of human obligation to other organisms and the biosphere as a whole is best secured indirectly. Put differently, as I have elsewhere stated this, Jonas articulates an anthropocentric ethics qualified for the purpose of preserving the whole of natural life.

44. Ibid.

The inherent "indigence and fragility" of the organic constitutes it as a moral object for responsibility. Only what is living can be such an object, insofar as the intrinsic vulnerability of bios makes it necessarily a matter for care to those who possess the cognitive and affective capacities to recognize that vulnerability. Only humans have these capacities and thus only humans have the capacity for being morally responsible. While the horizon of moral objects extends to all that is animate, the plane of agency is limited to human life. All biotic forms are moral objects, but only humans can be moral agents. Following this, Jonas argues, "Man's distinction that he alone can *have* responsibility means also that he *must* have it [first of all and principally] for others of his ilk—that is, for such that are themselves potential bearers of responsibility...."[45] Though the capacity for responsibility is unique to humans, it is the inherited outcome of natural evolutionary processes. Responsibility, in other words, is not a feat of human endeavor but a fact of human nature. It is simultaneously a mark of continuity and discontinuity with the rest of life. The fact that humans by nature *can* be responsible, Jonas suggests, grounds the claim that they *ought* so to be toward others who also *can* be. Thus Jonas's analysis inverts the Kantian dictum that "Thou canst because though ought." Jonas's formula suggests that, "thou ought because thou canst." Because responsibility is a defining capacity of human nature, because humans alone among organisms can be responsible, because this responsibility is the outcome of nature's intuitively good purposiveness, humans ought therefore to be responsible. Because humans alone among life's diverse forms have the cognitive leverage to imaginatively predict the possible consequences of their efficacy, and the affective capacity to respond caringly to those under their power, humans by nature are morally responsible. In this way, then, responsibility is ontologically grounded in the nature of human being. Jonas hereby circumnavigates the doctrinaire modern chasm between "is" and "ought". He writes, "an 'ought' is concretely given with the very existence of man; the mere property of being a causative subject involves of itself *objective* obligation in the form of external responsibility."[46]

---

45. Ibid., 99.
46. Ibid.

The first imperative from which all others follow, as Jonas articulates it, is the imperative to ensure the continued existence of the idea of humanity.[47] Humanity's continued existence has in all prior times been taken for granted, and was always implicit. But in a time of radical power in which it is possible to eradicate the human future, not to mention the entirety of the biosphere, this imperative needs to be raised to the level of the explicit. Further, the imperative to preserve the continued existence of the idea of humanity includes within it the imperative to preserve the necessary conditions for responsibility. This is so according to Jonas's view that the capacity for responsibility is a defining capacity of the human species. To ensure the future existence of humanity therefore implies the imperative to ensure the conditions of possibility for the exercise of moral responsibility. If the capacity for being responsible is part of the essence of human being, then exercising responsibility for the possibility of future human responsibility is necessary to the continuing possibility of genuine human being.

Though necessary to the future of humanity, is this imperative sufficient to the preservation of nature? Why ought the future of the

---

47. This is, as mentioned before, an admittedly confusing claim in an ethics as radically concerned with the natural environment as Jonas's, though I have argued that it is less confusing when understood as an aspect of Jonas's "qualified anthropocentrism". Earlier I cited Richard Bernstein's critique on this score. Lawrence Vogel, as appreciative of Jonas's work as Bernstein, raises what he terms a "psychological" critique of Jonas's privileging of the idea of humanity. It is worth mentioning here. He suggests, "One might wonder, along with all advocates of 'the hermeneutics of suspicion,' whether the insistence that we be subject to an imperative coming from beyond us lest 'things fall apart' is not driven by a psychological need to be at the center, linked to eternity, in order to avoid the bitter truth of our contingency. There is no hint of the terror of nature in Jonas's writings." See Vogel, "Does Environmental Ethics Need a Metaphysical Grounding?" With respect to the first, hypothetical part of this critique, whether or not Jonas's imperative may emerge in some ways from psychological need, I think it clearly does. Jonas's project stems in significant measure from his own life experiences, his sensed and experienced encounters with vulnerability, which of course include a psychic dimension. Jonas himself acknowledged this and I am not sure how it compromises his ethics. His concern for life, human and natural, arises from a deeply felt sense of the vulnerability of life. With respect to the second aspect of this critique, that Jonas does not acknowledge the terror of nature, I think Vogel is partly right. Jonas's primary concern is with the terrible ambiguity of human power. But at the same time, the peculiar capacities of human efficacy for impacting the world have themselves arisen through the processes of biological evolution. The human is part of nature, and human power can be terrifying, and to this extent Jonas does attend to at least one dimension of the terror of nature.

idea of humanity to be preserved? What justifies human existence? Considering the enormous impact of humans on the ecological viability of the biosphere, might the biosphere be better off without humans? Of course human history is littered with tragedy and atrocity, with humans behaving badly toward themselves and other forms of life. Yet Jonas insists, our primary responsibility is to the possibility of there being future responsible humans. This is not so because humans have always in fact exercised their power responsibly, for it is much too clear that this is not the case. Nevertheless, Jonas argues, the self-purposes of other organisms can best be guarded indirectly through the imperative that we ought never to do what might destroy the conditions necessary to the possibility of there being future generations of responsible humans. In Jonas's words, "the possibility of there being responsibility in the world, which is bound to the existence of men, is of all objects of responsibility the first."[48] But how can this be the case, given the human propensity to leverage short-term human desires, goods or rights over long-term ones, not to mention the momentum of the historical legacy of sacrificing nature for instrumental human purposes or desires? Has Jonas in his attempt objectively to ground ethics in an ontology of human nature compromised the security of nature's future? Turning once again to the evolutionary principle of organic continuity, Jonas answers an emphatic "No" to these questions.

The duty to preserve the future conditions of human responsibility includes within it, on the principle of evolutionary continuity, a duty to preserve the continuing integrity of nature. "Care for the future of mankind is the overruling duty of collective human action in the age of a technical civilization," writes Jonas, and this "care must obviously include care for the future of all nature on this planet as a necessary condition for man's own."[49] The duty to preserve the human future must include the duty to preserve nature, since apart from certain propitious natural conditions humans could not have become and cannot remain human. While on first glance, this care for nature may appear intractably anthropocentric, Jonas argues that it is not. For the capacity for responsibility that is the object of our primary imperative itself arises through natural processes. Humans are products of nature. In a startling

48. Ibid.
49. Ibid., 136.

inversion, then, the duty to preserve a future for human responsibility, which derives from respect for the human capacity for responsibility, itself derives from a deep appreciation for nature's intrinsic goodness.

The good of humanity and of nature are both subsumed in the imperative to preserve a truly human future. The imperative to preserve a genuine human future includes within it "a duty toward nature as both a condition of [human] survival and an integral complement of unstunted [human] being."[50] Not only are moral obligations to nature included within the imperative of responsibility, but also the intertwining of nature and humanity can also lead humanity to "rediscover nature's own dignity" and may command "us to care for her integrity over and above [this] utilitarian aspect."[51] And yet a problematic fact remains—though the offspring of nature, humanity has arisen to become the possible cause of nature's own demise. In Jonas's words, "Nature could not have incurred a greater hazard than to produce man.... In man, nature has disturbed herself...."[52] This claim constitutes a generative paradox in Jonas's ethics. The paradox is that the human for Jonas is both an effect of creative natural processes and also, especially in the present, a possible cause of nature's destruction. The generativity of this paradox pivots on Jonas's claim that it is upon recognizing nature's vulnerability in the face of the threat of radical human power, combined with an intuition of being's goodness, that moral responsibility gains traction and that an appreciation for nature's goodness can arise.

And yet, in spite of the claim that human moral obligation to the natural world and for the future of the idea of humanity has an intuitive basis, Jonas is not overly optimistic about the capacity of present human communities to embody his vision of responsibility. The practical demands of Jonas's imperative are radical, and he deems political triage to be requisite to its enactment. Concentrated, centralized political power is needed to counter the present powers of humanity that threaten the possibility of the responsible exercise of power upon which the future of humanity and the whole of life depend. In the present, "we are ... confronted with a dialectic of power which can only be overcome by a

---

50. Ibid., 137.
51. Ibid.
52. Ibid., 138.

further degree of power itself, not by a quietist renunciation of power."[53] The threatening dialectic of power here referenced is the concatenation of the forces of exponential population growth and the demands that growth places on finite natural resources, along with the always morally double-edged force of technological innovation.

In Jonas's judgment, the political and economic successes of liberal democracy and free-market capitalism—in which efficiency and increase of production fuel the consumption patterns of an ever-expanding free global citizenry that claims as its natural right access to and acquisition of more and more goods—combine "enormously [to increase the] metabolism of the social body."[54] The Malthusian law of populations, in which no individual species can grow indefinitely without ultimately undermining the conditions of its survival, has been "bullied to the extreme of [its] tolerance."[55] The power needed to counter the power of humanity, in order to save the future of life from humanity, however, must arise from within human society. But time is not on our side, in Jonas's view. The luxury of time required for the formation of an ecologically enlightened civic culture, for the "grass-roots" greening of democracy, is no longer available. Instead, "only a maximum of politically imposed social discipline can ensure the subordination of present advantages to the long-term exigencies of the future."[56] The urgency of a time of radical power and the rigorous demands of the imperative of responsibility require imposition of policy and regulations from a centralized government.

Jonas is not naïve about the prospects of a new socialism, however. He is emphatically anti-utopian, as this chapter earlier claimed. History has shown that the arguably laudable theoretical advantages of political centralization have in practice significant liabilities. Virtuous governance from the top requires a political culture below that nourishes the positive formation of political character. But this requirement is compromised by the historical evidence that centralization tends to result, among other things, in the atrophy of political will and moral initiative on the ground. Further, economically, the socialist embargo

---

53. Ibid., 141.
54. Ibid., 140.
55. Ibid., 141.
56. Ibid., 142.

on market competition discourages thrift and efficiency in production. Nevertheless, Jonas argues, the need-economy of socialism is a better model for our time than the profit-economy of capitalism. More importantly for Jonas, considering his view of the emergency-state of the present situation, socialism includes the added benefit over democracy of possessing the power to impose the necessary but likely unpopular regulations that the imperative of responsibility entails. While democracy can only hope for the cultivation in the present of a self-restraining populace, socialism can enforce this. And, as stated earlier, Jonas does not consider the current historical moment as one in which hope should be a guide. In light of this, he calls for a critical-retrieval of socialism, a greening of the red. The power over power that makes the socialist paradigm at least temporarily attractive requires that it reinterpret "its role from bringer of consummation to preventer of disaster, that is, by renouncing its breath of life—Utopia."

A paradoxical relation between vulnerability and responsibility is at the core of Jonas's ethics, as my earlier exegesis indicated. And a paradoxical relation between popular and centralized power is at the center of his anti-utopian political prescription. I will return to this paradox in my concluding chapter, in which I will argue that, in its relative neglect of the significance of moral formation to an ethics of responsibility, it is one of the principal weaknesses of Jonas's ethical vision.[57] But now it is time to turn to what I interpret as a paradox within Jonas's theological vision, thus far unexamined, that though he deems theology a "luxury of reason," theology plays a vital role within his project. In what follows,

---

57. I will suggest, in fact, that Jonas's political prescription actually undermines the formation of responsibility upon with the future of life depends. His call for a centralized government atrophies the political will of the many to take initiative on their own to preserve nature. Richard Bernstein's article, "Rethinking Responsibility," is instructive on this score. I will refer later to this. A stinging critique of Jonas's politics is also supplied by Ullrich Melle, from which I quote at length: "[Jonas] seems to think in terms of a very crude opposition between the masses of people who are driven by their insatiable desire for more wealth, comfort, and individual liberties and an elite which rules over them. And it seems that he can think of collective responsibility only in terms of such an elite as a group of guardians who, enlightened by [his own] ethics of responsibility, take the necessary measures and, if nothing else avails, even resort to deception and lies, to have the necessary restraints accepted by the masses." Though hyperbolic, and in my judgment completely misunderstanding Jonas's theological myth as strategic "deception", I take Melle's basic critique of Jonas's political dichotomy to be a suggestive one. See Melle, "Responsibility and the Crisis of Technological Civilization," 336.

I will critically interpret what Jonas means by calling theology a "luxury of reason" in order to prepare the way for the next chapter's critical comparison with Gustafson's theocecentric ethics, and, ultimately, to suggest the way in which Jonas's theological speculations provide a useful complement to Gustafson's and a helpful platform upon which to develop the theological aspect of an anthropology of responsible participation.

The Role of Theology

Thus far this chapter has interpreted several core features of Jonas's ethics. I recounted his view of the importance of an ethics for the future in the contemporary situation. I summarized the articulation of his imperative of responsibility and then examined his argument for its ontological grounding. I exegeted his analytic of responsibility, the parental and political models of nonreciprocal responsibility, and briefly discussed his political vision just above. It is important to reiterate that much of what has been interpreted issues from Jonas's motivation to articulate a conception of the good robust enough to constrain radical human power. As my previous chapters show, he takes metaphysical and ethical dualism, rooted in Gnosticism, and what he views as their inevitable consequence of nihilism, to be his primary challenges. In effect, one of his aims is to convince the nihilist that there is a good beyond the will-to-power, objectively grounded in the nature of things, to which all are held responsible. His theological writings, though he does not deem them to be integral to his task, are continuous with this aim.

I interpret Jonas to mean three things in his claim that theology is a "luxury of reason". First, he deems that the idea of God as author or creator of the world is unnecessary to perception of nature's goodness. That nature is good, for Jonas, is intuitively evident apart from faith in God. Second, the existence of God is unnecessary to the ontological grounding of goodness. For Jonas the question of "why there ought to be anything rather than nothing" is not adequately answered in theological terms. Answering "God" to this question just regressively extends the question. Third, motivation and enactment of the imperative of responsibility does not require the sanction of divine threat or reward.

If this is the case, the obvious question arises as to what purpose an interpretation of Jonas's theological writings serves. Though Jonas does not view God as a requirement of reason, the existence of God can neither be proved nor disproved, and he has speculated as a philosopher about the possibility and meaning of such existence. Faith is not necessarily unreasonable for Jonas. Indeed, in his essay, "Matter, Mind, and Creation," he speculates that though beyond reason to prove, the existence of God is plausible.[58] He argues that since the inwardness and intentionality characteristic of human life comes forth through material substance, such characteristics cannot be utterly foreign to the structure of matter. If they cannot be foreign to matter's structure, they cannot be foreign to the origin of matter. Given these claims, subjectivity must be a latent possibility at the very beginning of the universe. As he puts this, "[if] matter from the very beginning is mind asleep, so we must immediately add that the really first cause, the creative cause, of mind asleep can only be mind awake."[59] The existence of human life, then, leads to the plausibility of a "mental, thinking, transcendent, supertemporal being at the origin of things."[60] For Jonas, then, the existence of God is plausible even if it cannot be rationally proved.[61]

Thus discussing Jonas's theological writings is relevant, first, since his corpus includes, and even culminates in theological work. Second, Jonas's theological writings respond to two of the same problems that drive his philosophical ethics: the problems of reason and evil. Recall that the problem of reason for ethics, in Jonas's judgment, is that in its scientific materialist orientation it has undermined the articulation and grounding of values. The problem of evil is related to this for Jonas insofar as it concerns the unmooring of power from limiting objective normative constraints. Looking to his theological response to these problems lends insight into his larger philosophical ethical vi-

58. See Jonas, *Mortality and Morality*, 165–97. This essay was originally presented in abbreviated form as one of the opening lectures at the international conference "Mind and Nature," held in Hannover in 1988.

59. Ibid., 181.

60. Ibid., 182.

61. And this is my point in citing this essay and argument. I of course do not mean to suggest that I take Jonas definitively to have answered the question of God's existence, only that he takes such existence to be plausible within the bounds of modern cosmological thinking. This in itself is a significant claim to note about a thinker such as Jonas whose task is to articulate a naturalistic grounding for ethics.

sion. Third, a discussion of Jonas's theology provides a necessary point of critical comparison with Gustafson's intensely theological vision. Fourth, and most importantly, it may be that theology is vital to Jonas's project, his own claims to the contrary. Indeed, in what follows I will suggest that his theology is integral to his project, and specifically to his moral anthropology.

Jonas's theological response to the problem of reason is developed principally in three essays, "Immortality and the Modern Temper," "Is Faith Still Possible?: Memories of Rudolph Bultmann and Reflections on the Philosophical Aspects of His Work," and "Matter, Mind, and Creation: Cosmological Evidence and Cosmogonic Speculation." Jonas's common burden in these writings is to argue for the compatibility, though again not the necessity, of certain theological convictions with modern scientific knowledge. I will focus my interpretation on "Immortality and the Modern Temper," for in addition to illustrating how he identifies compatibility between the theological idea of immortality and modern knowledge, it is here that he presents a methodological justification for, and I will argue, the crucial importance of theology to his ethics.[62]

Jonas's comments at the opening of this essay are especially instructive. He distinguishes between the *object* of the idea, immortality itself, and the *idea* of immortality. Immortality as object is not available to knowledge, open neither to proof nor disproof. In contrast, the idea of immortality is a subject of knowledge, and "the intrinsic merits of its meaning become the sole measure of its credibility, and the appeal of that meaning remains as the sole ground of possible belief—as certainly the lack of such an appeal is sufficient round for actual disbelief."[63] As I will show soon, I take this remark to reflect a moral approach to the question of immortality. The grounds for belief or disbelief are not theoretical or logical, but practical; the question of the validity of the idea of immortality turns, for Jonas, on its the practical-moral merits. Extrapolated, Jonas's approach to immortality in this essay is suggestive of his broader understanding of the role of theology in ethics. In

---

62. "Immortality and the Modern Temper" was originally delivered as the 1961 Ingersoll Lecture at Harvard Divinity School. It is included as the fifth chapter in *Mortality and Morality*. Hereafter the page numbers will refer to the essay in *Mortality and Morality*.

63. Jonas, *Mortality and Morality*, 115.

a Kantian vein, his approach is that of an ethico-theologian. Also in his opening comments, Jonas stipulates that the point of his essay is not so much to say anything new about a topic that has been endlessly discussed, but to treat it as a portal for understanding "the present state [of human thinking about] our mortal condition."[64] In a way that is similar to my interpretation of his early work on Gnosticism, Jonas's concern with the ancient problem of the idea of immortality is a contemporary one—the current state of thinking about the meaning and value of finite human life.

In focusing his inquiry on the *idea* or meaning rather than the *object* of immortality, Jonas centers on what he takes to be the two principal issues at stake in the idea: justice, and the distinction between the phenomenality and reality of time. Both of these topics, in their usual relation to the idea of immortality, find resistance in the "modern temper," and yet it is this very resistance for Jonas that provides hints for the enduring merit and moral significance of the idea of immortality. Jonas argues that the conviction that justice waits upon a future time is viewed by the modern mind with repugnance. Temporal misdeeds require temporal retribution, and the idea that undeserved suffering requires future reward is held in doubt by the contemporary questioning of the rightful claim to happiness to begin with. "Indeed," he writes, "the here cannot be traded for a there—such is our present stance."[65] This emphasis on the temporal here and now is the source as well for the contemporary resistance to distinguishing between appearance and reality, "of which the idea of the mere phenomenality of time is a case."[66] The idea of an eternal order in which human immortality participates, runs, on Jonas's reading, radically counter to contemporary historical consciousness: "From the discovery of man's basic historicity to the ontological elaboration of the innermost temporality of his being, it has been borne in upon us that time, far from being a mere form of phenomena, is of the essence of such things as selves, and that its finitude for each is integral to the very authenticity of his existing."[67] Historical

---

64. Ibid., 116.
65. Ibid., 118.
66. Ibid.
67. Ibid., 119.

consciousness, then, is the source for Jonas of the general unresponsiveness of "the modern temper" to immortality.

And yet it is also this very historical caste of mind, for Jonas, which provides a way to forge a new appeal for immortality. For it is within time that humans recognize "that temporality cannot be the whole story,"[68] even if traditional ways of thinking of immortality and eternity may not serve as proper stories either. In keeping with the theme in his project that what is most precarious is most valuable, Jonas argues here that what may be "immortal" in human life is not what is essential or enduring, but what is fleeting. In place of the immortality of the soul, Jonas argues for the immortality of *deeds* in a way that supports the whole thrust of his ethics of responsibility. The human sense of a realm transcending time, he reasons, is most intensely recognized in moments of decision. "In moments of decision, when our whole being is involved, we feel as if acting under the eyes of eternity," and it is in these moments that "the responsible agent [is placed] between time and eternity."[69] This paradoxical relation between the fleeting moment of decision within the transience of time, and the "self-surpassing,"[70] transcendent feeling that sometimes accompanies it, hint for Jonas toward an understanding of immortality that is at once compatible with the modern historical caste of mind and with his interpretation of the non-literal meaning of traditional conceptions of immortality. He writes, "On the threshold of deed holding time in suspense, but not a respite from time, [the point of the moment of decision] exposes our being to the timeless and with the

---

68. Ibid.

69. Ibid., 120, 121. The use of the phrase "moments of decision" in this context again signals the way in which Jonas deploys but reinterprets an existentialist idiom. In contrast to, for example, Jean-Paul Sartre, while the "moments of decision" Jonas is referring to here are indeed related to acts, they are acts the consequence of which have to do with a great deal more than the project of an individual subject's historical existence. I am thinking here especially of Jean-Paul Sartre's famous claim, "It is only in our decisions that we are important." At stake for Jonas, and this will become more evident in my interpretation of his speculative theological myth that follows, the "important" existence at stake is not merely one's own but the inclusive existence of the whole of life, the future of humanity, and even the being of the divine. As was mentioned early on in chapter one on the influence of Heidegger in his thinking, Jonas's use of "existentialism" is radically extended, applied not just to individual human subjectivity, nor to inter-human social relations, but to the phenomenon of life in general.

70. Ibid., 119.

turn of decision speeds us into action and time."⁷¹ This understanding affirms the modern emphases on temporality and finitude, he argues, as well as the "transcendent dignity" traditional views of immortality imply with respect to human decision and action.

After interpreting the linkage between this modern account of immortality and select traditional religious metaphors by way of an existential emphasis on their common phenomenological moral meaning,⁷² Jonas speculatively suggests a theological myth into which a contemporary appreciation for the idea of immortality might have a place. This "tentative myth," he claims, he would like to believe to be true—"in the sense in which myth may happen to adumbrate a truth which of necessity is unknowable and even, in direct concepts, ineffable, yet which, by intimations to our deepest experience, lays claim upon our powers of giving indirect account of it in revocable, anthropomorphic images."⁷³ The myth he presents provides a storied—layered and narrated—framework for his vision of the ethics of responsibility. Its narrative layers suggest a way to conceive the compatibility of reason and faith, and this in a way that allows for talk about and faith in God in the midst of the reality of evil. In support of the use of this mythic framework, Jonas

---

71. Ibid., 121.

72. He relates it to the Jewish tradition of the "Book of Life" symbol and the Manichean tradition of an eternal image or transcendent alter ego. Of the "Book of Life" symbol, Jonas replaces the traditional view of the inscription of names according to merit with the idea that deeds themselves are inscribed. What is important in this revision is not that individual lives are immortalized, but that deeds themselves are what is eternally registered and thus become all the more crucial to the transcendent realm. He asks, "Might it not even be, to venture yet a step further, that what we thus add to the record is of surpassing import—not indeed for a future destiny of ours, but for the concern of that spiritual account itself kept by the unified memory of things—and that, although we mortal agents have no further stake in the immortality which our acts go to join, these acts of ours, and what though them we make of our lives, may just be the stake which an undetermined and vulnerable eternity has in us?" (Jonas, *Mortality and Morality*, 123). This reversal of vulnerability is also discussed in his account of the Manichean tradition, in which one's "image" or "double" is held to exist in an eternal realm and to be determined by one's acts and deeds in history. Jonas also recounts of this tradition that there is a less individualistic version in which the source of what is produced in eternity is not a single being but the whole of world process. In this case, "God's own destiny, his doing or undoing, is at stake in this universe to whose unknowing dealings he committed his substance, and man has become the eminent repository of this supreme ... trust" (Jonas, *Mortality and Morality*, 124).

73. Ibid., 127–28.

draws on Plato's recourse to myth in Book X of the *Republic* in which the construction and use of myth is advanced as a reasonable pedagogic device for symbolizing what is true but what lies beyond the reach of knowledge. Too, he follows Kant in viewing rational metaphysics not as excluding what is beyond knowledge, but as making room for faith.[74]

Jonas's myth is of course strongly shaped by Jewish monotheism, and, more particularly, the Rabbi Isaac Luria's Kabbalistic teaching of *tzimtzum*, or divine self-contraction. According to Jonas's myth, a revised account of the Lurianic version, God's power is completely exhausted through the act of creation. God totally divests and renounces his transcendent divinity in order that the world might be autonomous and for itself.[75] Jonas writes, "In the beginning, for unknowable reasons . . . the Divine chose to give itself over to the chance and risk and endless variety of becoming. [And God did this] in order that the world might be, and be for itself . . . ."[76] The evolutionary process through which matter comes to life, primitive life becomes animal and eventually human, is the gradual reawakening of the divine transcendence to itself. In the earliest stages of this development, prior to conscious life, God's cause, a good and free world, is safe. With the earliest forms of life, God only increasingly regains the self-experience initially surrendered. This regaining of divine experience turns for Jonas on the idea that life, in contrast to mere matter, is connected to mortality. He writes, "with the pitch of awareness heightened by the very press of finitude," the "divine landscape bursts into color and the deity comes to experience itself."[77] But then God "trembles as the thrust of evolution, carried by its own momentum, passes the threshold where innocence ceases . . . . The advent of man means the advent of knowledge and freedom, and with this supremely double-edged gift the innocence of the mere subject of self-fulfilling life has given way to the charge of responsibility under the

---

74. Lawrence Vogel makes this claim as well in his excellent editor's introduction to *Mortality and Morality*, which is entitled, "Hans Jonas's Exodus."

75. According to Vogel's essay noted above, Luria's account of God's self-contraction does not indicate the total eclipse of divine power. God's power is, so to speak, held in reserve. God still reserves the power to intervene. According to Jonas, this yet presents a problem with respect to the world's evil. For if God can intervene, but chooses out of power not to intervene, such a God is morally incomprehensible and unworthy of faith.

76. Jonas, *Mortality and Morality*, 125.

77. Ibid., 126.

disjunction of good and evil."[78] With the rise of knowledge and freedom in human life, and the opening of the morally ambiguous zone of increasingly mediated self- and world- relation, God's cause become precarious.

After presenting this myth, Jonas explicates some of its ethical implications. Essentially, the myth reinforces his account of the immortality of human deeds and the profound responsibility of human action. For human deeds within the myth attain transcendent causal and moral efficacy. According to the myth, the human is not created "in" but "for" the image of God, and human deeds are the source of either the "perfection" or "disfigurement" of the divine image.[79] In addition to this, Jonas suggests that his myth construes all existence as gift. Neither the world nor any creature in it has to be. The world came to be as it is only on account of God's self-denial, and to "this self-denial," Jonas writes, "every creature owes its existence."[80] God has nothing left to give, and the human task becomes that of giving being back to God.

It would seem that this would matter a great deal to the conduct of human life, but Jonas asks, in light of the world's evil and profound suf-

---

78. Ibid., 127.

79. Ibid., 128. This account of the *imago dei*, or in Jewish terms *tselem elohim*, is one that according to Hebrew Bible scholar Tikva Frymer-Kensky is in some ways distinctively Jewish. See "The Image: Religious Anthropology in Judaism and Christianity," in *Christianity in Jewish Terms*, 321–37. Frymer-Kensky does not specifically refer to Jonas, but in an illuminating essay on Jewish and Christian religious anthropologies, she notes both differences and commonalities. As I understand her account of this, Jonas's distinction between being in the image "for" rather than "of" God relates to the early rabbinic understanding of *tselem elohim* in *quantitative* terms, in contrast to the early Christian and especially the Pauline *qualitative* conception of the *imago dei*. As Jonas has presented this, the human was created "for" the image of God and thus the human moral and religious task is one that focuses on deeds or actions that can either increase or diminish the image of the divine in the world. Frymer-Kensky also notes of course that there are different traditions of this within Judaism and Christianity, that there is no singular "image" tradition in either religion. But she does note some fundamental understandings that underlie the differences within and between the traditions, the most basic of which is that the idea of the "image" underscores that for Jews and Christians alike "the image of God forms the basis of a religious anthropology that stresses the God-like aspects of human existence" (330). I will consider Jonas's understanding of *tselem elohim* more carefully in my concluding chapter, and attempt to point to ways in which it may provide a constructive complement to Gustafson's theocentrism and be useful to the development of a theological moral anthropology of responsible participation.

80. Ibid., 129.

fering, especially in light of the Holocaust, whether it really does matter. The question turns for Jonas on the status of those whose lives have been cut short, slaughtered, who for reasons of the massive cruelty of others' deeds did not experience the gift and responsibility of their own capacity for action. Jonas writes:

> I am thinking of the gassed and burnt children of Auschwitz, of the defaced, dehumanized phantoms of the camps, and of all the other, numberless victims of other man-made holocausts of our time.... [Are these] debarred from an immortality which even their tormentors and murderers could obtain because they could *act*—abominably, yet accountably, thus leaving their sinister mark on eternity's face?[81]

Jonas refuses to believe this, and chooses to believe instead that all human life stands accused by God's "weeping in the heights at the waste and despoilment of humanity . . . ."[82] In the face of the horror of the Holocaust, the magnitude of its evil, and the extension of morally ambiguous human power that defines the ethos of the present, the image of God is more precarious now than ever. The moral significance of human deeds in the present world of history and nature, according to this myth, is radicalized by way of finite human participation in a transcendent eternal realm in which the effects of acts never cease to reverberate and upon which Jonas's account of the divine cause, the unfolding freedom of the world's goodness, depends.

While in "Immortality and the Modern Temper" Jonas responds principally to the problem of reason in theology, in "The Concept of God after Auschwitz" he is more focused on the problem of evil. Jonas returns again to his myth in this essay, but emphasizes its theological over its ethical implications.[83] The pivotal question is: "What God could let the Holocaust happen?" The Holocaust for Jonas, and his own mythic construction, reveal a God "no longer assimilable by [some of] the old theological categories."[84] According to his myth, God is suffering, becoming, caring, and by way of these attributions *not omnipotent*.

81. Ibid.

82. Ibid.

83. As included in *Mortality and Morality*, this essay is Jonas's own translation of a lecture originally delivered in 1984 at Tübingen University. It also includes verbatim the myth used in the 1961 Ingersoll Lecture, "Immortality and the Modern Temper."

84. Jonas, *Mortality and Morality*, 133.

Jonas carefully distinguishes his account of divine suffering from the Christian theological view "of a special act by which the deity at one time, and for the special purpose of saving man, sends part of itself into a particular situation of suffering (the incarnation and crucifixion)."[85] In Jonas's myth, God suffers from the very moment of creation, all throughout creation, and with the life of every creature. This understanding, he argues, does not conflict with the biblical portrait of a God, who, on his reading, is shown to be a God that grieves and is affected by the world.

God's suffering relates to God's becoming and caring. That God is becoming rather than completed self-subsisting being may conflict with Hellenic concepts, but again for Jonas, not with the biblical imagery of God. To view God as becoming is to view God as affected by the world, truly in relation to it, which for Jonas is "the cardinal assumption of religion . . . ."[86] Along with a suffering and becoming God, the myth portrays God as caringly involved in the world. This is suggested by the initial divine divestment that allowed the world originally to come into being. That God is symbolized as caring is not an unfamiliar understanding of God. But Jonas's myth does not presume the fulfillment of God's care for the world, that all will be well. Rather, it imagines that God "has left something for other agents to do and thereby has made his care dependent on them. [This God] is therefore also an endangered God, a God who runs a risk."[87] It is up to humanity to bring to fulfillment God's care for the world.[88]

85. Ibid., 136. There is resonance here between Jonas's conception of the suffering God and Gustafson's christomorphism. As interpreted in chapter four, Gustafson does not hold to an objectivist Christology or to a satisfactionary theory of the atonement, but to a subjectivist, moral-influence Christology. Gustafson's emphasis, similar to Jonas's here, is not on a single event in history, but on the way in which the suffering love of God, demonstrated by Jesus, continues to motivate and judge human life and can be experienced in and through the whole of existence. That Jonas seems to equate an objectivist Christology with *the* Christian theological view evinces a rather unhistorical sense for the variety of Christian theological understandings of the significance of Christ.

86. Ibid., 137.

87. Ibid., 138.

88. This is a long way from Gustafson's theocentric vision. Differently from Gustafson's vision of God's purposes for the world, Jonas's myth suggests that the human is much more than a participant within the divine caring for the world, but is responsible for the fulfillment of this care. Crucially, Jonas holds that God does not

In combination, Jonas's mythic symbolization of divine attributes, and his concern for the evil of the Holocaust, lead him to reject the classical (Jewish and Christian) theological emphasis on God's omnipotence. Jonas expresses his rejection of God's absolute power in both logical and theological terms. Logically, according to Jonas, God cannot be all-powerful if there is anything else in existence that has power. That there are creatures in the world that effect it, that there are human actors, mean that God cannot be absolutely powerful. Absolute power is logically self-contradictory since the concept of power only has meaning "in relation to something that itself has power."[89] In making this argument, Jonas reflects the understanding of freedom theorized in his ethics and philosophical biology. He writes, "The situation is similar to that of freedom in the human realm; far from beginning where necessity ends, freedom consists of and lives in pitting itself against necessity. Separated from it, freedom loses its object and becomes as void as force without resistance."[90] In other words, as Jonas understands that freedom is always to some degree needful, so also power only has meaning in relation to a powerful counterpoint. On theological terms, Jonas argues that omnipotence must be rejected to save the intelligibility and goodness of God. To hold benevolence and omnipotence together inevitably redounds for him to a completely inscrutable, radically "hidden" God. After Auschwitz, God is benevolent, omnipotent, utterly unintelligible in cruelty, and thus unworthy of any kind of religious devotion, or good, suffering, caring, becoming, and impotent. In a reversal of the theodicy of Job, Jonas's response to evil is not to invoke "the plentitude of God's

---

guarantee the fulfillment of his care for the world, that all may not be well, but indicates, and I will pursue this further along, that God's cause for the world depends on the fulfillment of the human good—being responsible and ensuring the continuing possibility of human responsibility. This is in marked contrast with Gustafson's theological and ethical critique of the conflation of the human good and divine purposes, and is a contrast that the different anthropological metaphors of participation and responsibility bring into high relief. As I will attempt to argue in the next chapter and in my conclusion, Jonas's view is important to consider in light of the urgency of the environmental crisis and may provide helpful ballast to Gustafson's admirably pious, but perhaps insufficient normative counsel. The task will be to determine whether Jonas's position is in fundamental conflict with Gustafson's theocentrism or if there may be resources in Gustafson's project for incorporating Jonas's insight.

89. Ibid., 139.
90. Ibid., 138.

power" but its "voidance."[91] And yet, in conclusion to this essay, Jonas writes, "All this, let it be said . . . is but stammering."[92] But is it? I conclude this chapter by considering what Jonas's purported "stammering" in these essays reveals about his view of the relation between theology and ethics in contrast to Gustafson's, and what this suggests for my constructive purposes.

Conclusion

First and foremost, as reiterated numerous times, Jonas claims that his account of the objective good-in-itself does not depend on theological support. The objectively grounded and categorically binding character of the good is evident for Jonas within the nature of things, revealed through an existential interpretation of biological facts, and makes an intuitive moral demand upon the conscious will of human beings. Theology does not for Jonas meet a need of reason. And yet, at the same time, his ethics and methodological commitments do not rule out the idea of a Creator God.[93] Belief in God may not meet a need of reason, but neither is it implausible. Jonas's speculative theological writings, and the needs I will argue that they do serve, are leveraged on this claim.

Jonas's rather minimalist view of the plausibility of a Creator God, I argue, serves as a pivot for understanding Jonas's maximally imaginative, anthropologically vital theological speculations. It allows Jonas, through myth, to talk about the mysterious plausibility of God in non-dogmatic terms, without reducing God to a projection of human experience. Of this, Jonas writes, "The paradoxical sphere of divinity is better protected by myth, whose manifest opacity remains transparent for the ineffable and mysterious God, than by concepts grounded in the self-experience of man."[94] This statement is a long way from Gustafson's understanding that theological discourse should be rooted in human experience, and that the intelligibility and persuasiveness of the idea of God depends in large measure on what the sciences can validate.

91. Ibid., 142.

92. Ibid.

93. Lawrence Vogel writes, "Jonas contends that although we can make ethical sense of our place in nature without appealing to a transcendent Creator, we can also make sense of nature—and perhaps deepen its meaning—by thinking of it *as* God's creation" (Vogel, "Jewish Philosophies After Heidegger," 127 [italics original]).

94. Jonas, *Mortality and Morality*, 134.

Jonas reverses this logic. For him, it is myth rather than a theological discourse congruent with the sciences that preserves God from being reduced to human needs and desires. This is an inversion of Gustafson's argument that the sciences and human experience provide the surest filters against the instrumentalization of God. The significance of these differences between Jonas and Gustafson's theological methodologies provokes the question of why Jonas, who has endeavored so fervently to present, in his own terms, a purely philosophical, naturalistically grounded ethics, is concerned at all with preserving the mystery of God.

Answering this question, I think, can be aided by taking a brief biographical turn. Jonas's theological writings are deeply personal, grounded in the need and desire to make moral sense of his experience as a Jewish philosopher living after Auschwitz. It is at least in part from this need and desire, I argue, that his theological speculations arise. In spite of his naturalistic sensibility and his insistent self-identification as a philosopher rather than a theologian, the question of God never lets go of Jonas. Further, the existential vitality of Jonas's theological writings, I suggest, indicate their importance to his moral anthropology, even while they may not be necessary to his naturalistic metaphysics. This is most evident in "God After Auschwitz." The occasion for the essay was a reception ceremony for the 1984 Dr. Leopold Lucas Award granted to Jonas by Tübingen University. He chose to speak on his topic upon recognition that Rabbi Lucas had died in Theresienstadt and that Lucas's wife had shared the same murderous fate of his own mother in Auschwitz. Upon realizing this connection, he believed he "owed it to those shadows that something like an answer to their long-gone cry to a silent God not be denied to them."[95] As interpreted earlier, Jonas endeavors in this essay to provide an image of God that is morally comprehensible in the midst of the reality of evil. This moral intelligibility of God is one of Jonas's primary themes in his theological writings.[96] An

---

95. Ibid., 131.

96. In relation to this, Jonas writes, "The *Deus absconditus*, the hidden God (not to speak of an absurd God) is a profoundly un-Jewish conception. Our teaching, the Torah, rests on the premise and insists that we can understand God, not completely, to be sure, but something of him—of his will, intentions, and even nature—because he has told us. . . . A completely hidden God is not an acceptable concept by Jewish norms," (Ibid., 140).

omniopotent God who allowed the Holocaust to happen, who did not intervene to prevent it, is unintelligible to Jonas and unworthy of faith. To preserve the intelligibility of God and faith in the goodness of God, God's omnipotence has to be rejected. These are the speculative conclusions Jonas expresses through his mythic construction, motivated by his burden to answer not only the cries of those who were murdered during the Holocaust, but also the need within his moral anthropology of responsibility.

Here I part ways with Jonas's insightful interpreter Lawrence Vogel who suggests that while the self-divesting image of God evoked by Jonas's myth "offers one way of coping with the problem of evil, it would appear that this God is so self-effacing that we are left with a functional equivalent of a naturalistic view."[97] If this is the case, Vogel implies, Jonas's idea of God does not seem to be in any way responsive to the cries of the murdered. Rather than interpreting this as a point of critique, I take this to be precisely Jonas's point. Jonas's essay is emphatically not an attempt to speak for God, to provide a defense for God against the horrors of history. Jonas's theology strictly serves an ethical purpose, to deepen the meaning of his moral anthropology and to provide a mythic framework of support and motivation to his responsibility ethics. Jonas's God has no more to give to the world. The moral task of humanity becomes that of giving to the world what God no longer can: "And he may give by seeing to it in the ways of his life that it does not happen or happen too often, and not on his account, that it 'repented the Lord' to have made the world."[98] Jonas's response to the shadows of the murdered is to locate responsibility for evil squarely within the human moral adventure. His theological speculations call humanity to responsibility, not God.

On my interpretation, then, Jonas's theological writings, and especially his myth, reveal that theology is more for him than a luxury of reason. Though theology may not serve for Jonas a need of reason, in the sense of providing a ground for the good or motivation to responsibility, his theological writings nonetheless serve a vital purpose tied into his moral anthropology. Jonas's theology provides a deeper ground

---

97. From Vogel's editor's introduction to *Mortality and Morality*, 26. Put differently, the image of God in Jonas's myth is so self-effacing that it may make relationship with God impossible.

98. Jonas, *Mortality and Morality*, 142. Jonas here quotes from Gen 6:6–7.

for the good than intuition. Jonas's theological writings reinforce the gravity of the human moral task by situating responsibility for the good of nature and the divine cause entirely within the human enterprise. In sum, taking a cue from his own methodological commitment to the testimony of existential experience, I read the considerable attention Jonas grants to theology, especially in his later writings, as a hint that though he claims that the goodness of being is self-evident, he cannot let go of the question of God and is not satisfied that intuition is support enough for responsibility to the future idea of genuine humanity. He recognizes that while reason may be all that is necessary to the articulation of an ethics of responsibility, reason alone neither motivates responsibility nor exhaustively explains the yearnings for the transcendent immanent within the nature of human life.

In my concluding chapter, I will extend this argument by suggesting that the anthropologically vital role of Jonas's theology is further implied in his account of image-making. Recall that for Jonas, image-making, along with the tool and grave, "reveal various decisive human qualities," which, "taken together . . . provide us with something approaching essential coordinates of a philosophical anthropology."[99] Reflection on these artifacts, and their potential adumbration into various domains of human life and thought in the world, is required if a comprehensive anthropology is to be envisioned. For they are nature's own record of the human drive for transcendence, the pinnacle of the principle of mediacy that is for Jonas the interpretive key to the moral depths of bios: "with the tool [the human] surpasses physical necessity through invention; with the image, passive perception through representation and imagination; with the tomb, inescapable death through faith and piety . . . ."[100] Despite the enticement of this last quotation, though, it is now time to return to Gustafson, in order to bring this comparison with Jonas to a more fruitful conclusion.

---

99. Jonas, *Mortality and Morality*, 78.
100. Ibid., 85.

6

Theocentric Ethical Participation

Introduction

ONE OF THE SIGNIFICANT METHODOLOGICAL CLAIMS BEING ADVANCED in this book concerns the reflexive relation between description and prescription. Normative claims are inevitably shaped by a descriptive apparatus, which is developed in relation to interpretive and methodological choices, which themselves are influenced, always to some degree, by value commitments. The ways the world and the human are descriptively construed, and in a theological system, the way the idea of God is understood, support normative proposals and reflect evaluative-descriptive commitments. The indicative influences the limits and possibilities of the imperative and every indicative reflects a hermeneutic, which, as I earlier described it, I take to be partly chosen as a set of methodological commitments and partly given by a thinker's experience in the world and the traditions and histories that have impacted him. The rhetorical structure of this book reflects the tangled interactions of these several ethical methodological issues and advances attention to them as a critical task in the interpretation of ethical theories.

As should be clear by this point, Gustafson is a highly conscientious ethical methodologist—profoundly alert to the possibilities opened and the challenges created for any prescriptive claims by their relation to a descriptive context. The chapter on the hermeneutical dimension of his ethical theory indicated his commitment to the priority of experience, his view of experience as multidimensional, his understanding of the tasks of second-order ethical and theological reflection on experience, and his articulation of a revised Reformed theological framework. Together, these commitments reflect a hermeneutical position I earlier denominated as a critical religious naturalism—he attends methodologically to human experience in a complex, natural historical world for the purpose of developing what it indicates for the enterprise of theological ethics. The chapter on his fundamental ideas detailed his basic accounts of God, the world, and human being, and critically examined the way these ideas were shaped by his critical religious naturalism. In sum, these chapters bring to relief the experiential basis of his methodology and that God as the center of value is the substantive core of his vision. With this work in hand, it is now possible to move with greater insight into a critical discussion of the normative dimension of his project.

In what immediately follows, I will introduce Gustafson's theocentric ethics by situating it in terms of the several theological and philosophical traditions and thinkers that I take to be his primary interlocutors. The aim here is not to address all of the thinkers that Gustafson engages, but the ones that in my judgment best bring to view the distinctiveness of his project. In the second section, I will exegete his understanding of the basic theocentric moral question and the norm that flows from it, as well as what he takes to be the basic human flaw and its correction. And I will also specify the central objective and subjective dimensions of his ethics, what he takes to be the ground of his normative vision and the posture or form of life correlated to it. In my closing section, I suggest that as Gustafson's norm and anthropology of participation stand, they are not fully appropriate to a time of environmental crisis, and that his own methodological commitments justify revision to it.[1]

---

1. The task of critiquing Gustafson is enormously difficult. Jeffrey Stout goes as far as to say that, "The difficulty [of critiquing Gustafson], who always writes with an eye toward possible objections, [is that he] rarely misses an opportunity to qualify a thesis. The cumulative effect of this habit, which has served him well in the strictly analytical work that established his reputation as a leading living American interpreter of Christian ethics, is to deprive his *magnum opus* of much of its lucidity. [*Ethics from a Theocentric Perspective*] is a jeremiad in the great tradition, but one written in the cramped hand of an excruciatingly careful scholar." See Stout, *Ethics after Babel*, 173. I agree with Stout that Gustafson is an excruciatingly careful scholarly craftsman, but understand him to have presented an evocative perspective on theocentric ethics and to have invited other scholars to apprentice with him in its continued development.

Situating Theocentrism

While there are points of contact between Gustafson's theological ethics and other philosophical and theological ethical theories, his account of the distinctiveness of his own vision is that he takes to its necessary conclusions the conviction that God is the axiological center of moral life. If God is not "the guarantor of human benefits," as his understanding of God and his attention to the natural sciences leads him to claim, then theocentric ethics "may not be recognizable as ethics in terms of the most common understandings of it" in the traditions of the West.[2] For the most common understandings of philosophical and theological ethics feature the human, or an anthropomorphic image of God, as the center of moral reference. Thus while his theocentric ethics may resonate with certain features of other philosophical and theological ethical traditions, it is "religiously and morally uncomfortable" insofar as "the relating of all things in a manner appropriate to their relations to God does not guarantee benefits to [humans] as we traditionally perceive such benefits . . . ."[3] The moral sphere of Gustafson's theocentric ethics extends beyond the realm of human and even human-like subjectivity, purposes and values. According to his account of this, the will of God and the good of humans do not necessarily coincide.

These points require a profound recontextualization of the moral character of human existence and activity. The moral task is presented by Gustafson as the pious effort to coordinate character and behavior with the will of God, "a will larger and more comprehensive than an intention for the salvation and well-being of our species, and certainly of individual members of our species."[4] Further complicating this task is that in the condition of finitude and dependence, discernment can never achieve absolute certainty about what such coordination entails.[5]

---

2. Gustafson, *Ethics from a Theocentric Perspective*, 1:112.

3. Ibid., 113.

4. Ibid.

5. For Gustafson, as I will continue to explain, the moral and religious life of piety is inherently conflicted, intractably ambiguous, and often tragic. He writes, "Ethics in the theocentric perspective developed in this book does not provide absolute moral certainty or eliminate tragedy" (*Ethics from a Theocentric Perspective*, 1:342). Some of the reasons for this are that, "Limits of knowledge, foreknowledge, and the capacity to control many of the consequences of our interventions prohibit the achievement of absolute certainty" (*Ethics from a Theocentric Perspective*, 1:341). Human finitude and

In what follows, I will interpret the salient points in Gustafson's account of the relation of his theological ethics to other theories in order to prepare the way for my next section's more isolated exegetical focus on the structure and norm of his vision of the theocentric moral life.

The central difference for Gustafson between theological and philosophical ethics is that theological ethics begins from some account of God and God's interaction with the world. Such an account presents itself as the way things really are. Philosophical ethical systems too include convictions about the nature of reality, but do not present these in strictly theological terms. Though not beginning from explicitly theological premises, philosophical ethical constructions of reality may include a dimension of ultimacy that is functionally equivalent to theological construals of reality. According to Gustafson, however, this does not mean that it is accurate or fair to construe philosophical ethical systems as implicitly "theological." Rough equivalence means instead that interpretations of philosophical ethical systems can be potentially coordinated with theological ones. While it is inaccurate to interpret philosophical ethical convictions about ultimately reality as theological, theological ethics cannot on Gustafson's account avoid being in some sense philosophical. As Gustafson claims, even a theologian such as Karl Barth who infamously asserts that, "ethics is sin" makes judgments "that are the same as, or similar to, judgments made by moral philosophers."[6] Thus while the bases of theological and philosophical ethics may differ, comparison of theological and philosophical ethics is not only a possibility but also a possible means to mutual enrichment. Gustafson embraces these comparative and constructive possibilities especially through interpretations of utilitarianism and Kant's ethics on the one hand, and Karl Barth and Thomas Aquinas on the other.[7]

---

limitation impossibly complicate the effort to attain moral certainty. Related specifically to the context of the natural environment in the tasks of discernment, Gustafson suggests, and I opened ch. 2 with a discussion of this, that "We are a part of the natural world; it brings us into life and sustains life; it also creates suffering and pain and death" (*Ethics from a Theocentric Perspective*, 1:210). Given the simultaneously creative and destructive patterns of the natural world, the task of morally discerning what these patterns indicate for the life of piety is inevitably ambiguous.

6. Gustafson, *Ethics from a Theocentric Perspective*, 2:98.

7. Of course, as already noted, Gustafson's intense scholarly reflexivity means that his entire project is in conversation with multiple thinkers. I cannot interpret all of the thinkers he engages, but select here the ones that I take to be his primary interlocu-

Gustafson defends his focus of philosophical comparison on utilitarianism and Kant for two basic reasons: their enduring historical significance and the partial truths they embody. The key insights of the utilitarian tradition that Gustafson affirms include the prominence it grants to human purposiveness, its evaluative emphasis on the consequences of acts, and its empirical methodological commitment to human experience. But it is in relation to these points that Gustafson's theocentric ethics is also to be distinguished from utilitarianism. A helpful way to understand any ethical position is to consider how it responds to the basic moral question of what humans ought to do. For the utilitarian tradition, the general answer to this question is to do what yields the most morally good consequences. This of course leaves open a number of questions, as Gustafson is keen to point out, such as consequences for whom or what, measured by what standard or criterion? The general focus of attention in the utilitarian tradition, however, has been on purposive striving toward beneficial human consequences—whether construed by hedonists as the maximization of pleasure over pain, by preference utilitarians as the satisfaction of human preferences or desires, or by welfare utilitarians as the increase of human welfare.

---

tors and which in my judgment bring his ethical vision most fully into relief. This is in keeping with the aim of attempting to clarify Gustafson's project as much as this is possible in light of the way he presents it in such a densely analytic-comparative mode. For example, I do not grant considerable attention to H. Richard Niebuhr in this section, though Niebuhr was of course one of Gustafson's most significant mentors. I have elected not to provide a deep analysis of Niebuhr's influence for the simple reason that the strength of affinity between their ethical visions is not crucial to my aim of crystallizing the distinctiveness of Gustafson's. Having said this, however, I will comment briefly here, as I do further along in the text of this chapter proper, that I think Gustafson's ethics does alter his mentor's in some important ways. In contrast to Niebuhr, Gustafson strongly rejects to the use of agential language in symbolizing God. This relates to a second basic difference between Niebuhr and Gustafson. While affirming Niebuhr's radical monotheism, Gustafson's theocentric construal of God is significantly "higher" than Niebuhr's, at least in its epistemological and axiological respects. Jeffrey Stout offers suggestive analyses of these points in *Ethics after Babel*: "Gustafson wants to keep Niebuhr's radically theocentric orientation. What makes Gustafson's monotheism still more radical is the austerity of its God. We are simultaneously told that God's purposes should be the focal point of all ethical reflection and that we lack the means for bringing either God or his purposes into clear focus. Moreover, it remains unclear why, despite his stress on theocentrism, Gustafson insists on speaking of God or divine purposes at all?" (177–78). I will return to this perceived dilemma, here noted by Stout and earlier referenced in a different form by Gordon Kaufman, in the conclusion to this chapter.

Gustafson's theocentrism of course is concerned with a much broader scale of consequences, evaluated in relation to a nonanthropocentric standard.

While the varieties of utilitarianism do not necessarily neglect the larger wholes of social or natural contexts that frame their moral calculations, these wholes according to Gustafson are not granted more than instrumental axiological status. The wholes are construed as aggregates of individuals rather than independent entities with their own attendant intrinsic value. On Gustafson's interpretation, "any 'whole' that has to be taken into account [in utilitarianism] exists for the ends of its individual members."[8] Gustafson's theocentric ethics clearly departs from this idiocentric emphasis of utilitarianism, and yet he does not deny the moral significance of this departure. He acknowledges that one should "not move from the protection to individuals that [utilitarian] ethics sustains without good reasons."[9] For among other things, the ethical primacy of the individual has protected many from tyrannies of various sorts. Gustafson understands this and yet argues that from a theocentric perspective, the aspirations and obligations of human individuals are to be contextualized within and even sometimes sacrificed for the sake of the good of larger wholes.

Along with the utilitarian concern with consequences, Gustafson's ethics too is concerned with the effects of action. Among the difficulties endemic to any form of consequentialist ethics, though, are those concerning how broadly or narrowly to circumscribe the realm of moral significance, which metric or standard of value is characterized as being morally salient, and of how to predict consequences. These difficulties are only exacerbated in a time characterized by the radical expansion of human power. Gustafson's theocentric ethics shares with utilitarianism a concern with consequences, but certainly does not resolve the problem of prediction. In fact, by expanding the range of relevant consequences from individuals to larger wholes, even the whole of the cosmos, it only complicates this task.

This is not a point of critique in Gustafson's view but simply reflects the intractable ambiguity and tragedy of moral existence from a theocentric perspective. With respect to the range and metric of mor-

---

8. Gustafson, *Ethics from a Theocentric Perspective*, 2:109.
9. Ibid.,110.

ally salient consequences, utilitarian questions turn on some account of utility, or more generally, some view of what is constitutive of good or desirable effects. Various responses in the history of ethical thinking of course have been given and defended, such as human happiness, pleasure, or welfare. In contrast to these ideas, Gustafson's theocentric ethics, as he puts it, "has to deal with a variety of possible good effects and with the cost to other justifiable ends of achieving those selected."[10] As this quote dramatizes, the points of similarity and difference between utilitarianism and Gustafson's ethics underscore the degree to which theocentrism reflects an inevitably conflicted moral world.

Though Gustafson affirms the utilitarian tradition's attention to human desire and feeling, he critiques its basic anthropology for being too thin and egoistic. While attention to human experience certainly indicates concern to seek pleasure, avoid pain, and strive for happiness, it does not indicate that these are all that should be prescribed. Aside from the logical problem of the naturalistic fallacy here raised, of whether or how one can move from a descriptive claim about what humans desire and value to what they ought so to desire or value, there remains the moral problem of determining which ends or consequences are desirable or valuable and why. Gustafson's account of the interactive-participatory nature of human being means for him that more stress needs to be placed on the limitations of individual striving for happiness than he thinks utilitarianism provides.

The most significant difference between Gustafson's vision and utilitarianism, of course, is the axiological centrality of God in theocentric ethics: "All things are instrumental to the divine ordering, not to human happiness, and the divine ordering does not have human happiness as its final end."[11] Thus, in spite of affirming certain elements of the utilitarian tradition, Gustafson's view of the deep interpenetration of individuals and communities within the contexts of nature and human society, ultimately created and sustained by the divine governor of the universe, and the norm to relate to all things in a manner appropriate to their relations to God, clearly differentiate his ethics from utilitarianism in fundamental ways.

10. Ibid., 112.
11. Ibid., 108.

It is also crucial in Gustafson's judgment to situate his ethics in relation to Kant's, for no ethicist attempting to present a comprehensive vision can avoid engaging the purest expression of deontological ethics.[12] In what follows, as above, I will draw out the key points of Gustafson's critical interpretation of Kant in order to continue to specify the distinctiveness of his theocentric ethics. In contrast to the utilitarian concern with consequences of acts, Kant's moral focus is on the intention of agents. Kant is not unconcerned with effects of acts, but does not deem these as morally determinative. What determines whether an act is morally right or not is not the effects it yields but whether and how it conforms to moral principles and maxims. What is right is to act in and from duty to the moral law. Happiness or human benefit, in Kant's judgment, is a corollary of morality rather than its aim, incentive, or motivation. In fact, an act done for some benefit rather than in and from duty to the moral law, even if it may happen to conform to the moral law, does not count for Kant as a fully moral act. The intention driving conduct is crucial for Kant and the crucial intention, the only proper motivation or incentive to morality, is respect for the moral law.

At the core of Kant's moral theory is a rigid distinction between the moral and nonmoral realms. As earlier interpreted, Gustafson's theocentric ethics resists such a rigid distinction. Human moral capacities arise through natural processes and are shaped within social, cultural, and historical contexts. The ordering of nature for Gustafson is an important source for moral norms. But for Kant, the realm of morality is the realm of freedom, and freedom constitutes the human difference from the rest of nature. Without freedom there can be no moral law. Only an autonomous will can be held morally accountable for its acts. Though crucial, however, the reality of freedom cannot be proved. It is for Kant a necessary postulate of practical reason. In contradistinction to the realm of the moral, the realm of the nonmoral consists of nature or necessity. It is crucial to Kant's understanding of morality that necessity and freedom be separated. For to ground morality in anything

---

12. Along with Jonas, Gustafson is both appreciative and critical of Kant's moral philosophy. His basic concerns with Kant, roughly in keeping with Jonas's, are that Kant does not appreciate enough the moral significance of the natural and biological rootedness of human moral capacities by demarcating freedom and nature much too severely.

other than freedom is to ground it in contingency, and to be able only to posit hypothetical or prudential imperatives rather than categorical and ethical ones.

Gustafson affirms the objective grounding of Kant's ethics and the rigor of the moral life it advances, but he of course stakes out crucial points of difference. For one, Kant's ethics is in Gustafson's judgment radically anthropocentric. The sphere of the moral extends only so far as freedom, and since freedom belongs only to humans for Kant, ethical imperatives extend only to other humans. Though Kant does not reject the descriptive claim that humans exist within natural contexts and are biologically conditioned, Gustafson argues that he does not take such information to be morally significant. He writes that for Kant,

> the moral irrelevance of the linkage [between humans and nature] occurs because the moral has to do with only the transcendental freedom of the individual person. That which determines what is moral is the individual's autonomous will, the exercise of reason, and the capacity of reason, using respect for the moral law, to master and determine the phenomenal aspects of the self and (I take it) the natural world.[13]

In addition to critiquing the "mastery" of nature implied by Kant's ethics, Gustafson's understanding of the biological conditions of freedom and reasoning mean that the human relation to the natural world is highly relevant to any proper ethical theory. Since his moral theory is grounded in the noumenal realm of freedom, the patterns and processes of the natural world so crucial to Gustafson cannot for Kant "become a source of moral norms."[14] The descriptive distinction between the noumenal realm of freedom and the phenomenal realm of necessity that serves Kant's demarcation of the moral and the nonmoral is, on Gustafson's interpretation, strongly affected if not fully determined by his aim to defend an objective grounding for ethics and a rigorous account of morality. Though Gustafson acknowledges that Kant has reasons for making his judgment about this distinction, it is a judgment with which his ethics crucially differs. For Gustafson, the objective grounding of ethics cannot and should not sacrifice natural scientific

---

13. Gustafson, *Ethics from a Theocentric Perspective*, 2:131.
14. Ibid.

and common human experiential insight into the morally profound interrelation of humans in the broader natural world.

Another crucial difference between Kant's ethics and Gustafson's concerns the ways in which theology and morality are related.[15] For Kant, according to Gustafson, "the first intellectual task is to develop a theory of morality; what can be said about God is a corollary of that moral theory."[16] Thus in contrast to his own *very theological* ethics, in Gustafson's judgment Kant has a "*very moral* theology" and a very "untheocentric" understanding of religion.[17] Kant's and Gustafson's understandings of the relation between theology and ethics are basically inverted. For Kant, the method and content of ethics control what can be said and known about God; God is not a determinative but a regula-

---

15. The most basic theological difference between Gustafson and Kant reflects the most basic theological difference between Gustafson and Jonas as well. This difference can be thought of in terms of the distinction between physico- and ethico-theology that Kant makes in the appendix to the second division of his Third Critique, *The Critique of Judgment*. Roughly, according to Kant's formulation, a physico-theologian reasons empirically to the supreme cause of nature, God, and the properties or attributes of God from nature, while the ethico-theologian attempts to infer the cause of nature and the attributes of this cause from the moral purposes of rational beings. According to Kant, the resulting understanding of God for the physico-theologian is indeterminate, ethically inadequate, and merely propaedutic, while the ethico-theologian's approach leads to a more ethically adequate account of God. According to this formulation, though it is fair to say that Gustafson basically works as a physico-theologian and Jonas as an ethico-theologian, the ways in which they both challenge the Kantian demarcation of nature and freedom, or necessity and morality, reveal the insufficiency of Kant's theological typology for classifying them. For example, Jonas's existential methodology begins from phenomenological analysis of purposive beings in nature, but purposive beings are not on his reading exhausted by those who have the capacities of reason, and further, purposiveness is grounded in natural conditions. Thus while he moves from this analysis to a philosophy of nature and into to a constructive normative vision, and eventually to a speculative theological vision, to characterize him as a strictly ethico-theologian would entail overlooking the degree to which his accounts of the natural and the moral overlap and inform one another. Similarly, Gustafson understands that any account of nature's patterns and processes are partly influenced by preceding value commitments and the influence of historical and cultural traditions of thought, and thus that reasoning from nature to God is undertaken as a reflexive interpretation of the various interpenetrations of natural, historical, cultural, and social spheres of experience. Having indicated the ways in which both thinkers complicate the Kantian formulation, however, it remains the case that Jonas does *tend* in the direction of an ethico-theologian while Gustafson *tends* in the direction of a physico-theologian.

16. Gustafson, *Ethics from a Theocentric Perspective*, 2:137.

17. Ibid., 135.

tive moral idea. As with freedom, the ideas of God and immortality are necessary postulates of pure practical reason that cannot be scientifically proved or disproved. Yet they are ideas necessary to morality if happiness is, and it must be, truly a correlate of morality. For happiness obviously does not always attend moral rectitude in the present, and thus God and immortality must then exist to ensure the harmony of duty and reward. Morality is not to be motivated by reward, though in Kant's view it is in the end rewarded.

Gustafson acknowledges that Kant's reasoning about moral motivation is consistent with his own. One is not moral to be rewarded by anything in the present life or beyond, and yet duty to the moral law must have a reward. Though consistent, this reasoning in Gustafson's judgment further reflects the anthropocentric character of Kant's theory. He writes that though "part of what Kant reacted against is the mentality of 'God will get you for that' . . . his own justification of the idea of God comes close to being subject to the same charge." Kant can only be preserved from this charge in that "the assurance [of reward] comes after, rather than before, moral action, and thus is not a cause of it."[18] Gustafson strongly rejects the interpretation of divine reward and punishment as incentives to moral action, and so affirms Kant's own claims about this. Nevertheless, the practical postulation of God for Kant is necessary to ensure human benefit. In this sense, Gustafson charges Kant with instrumentalizing God. For Kant, argues Gustafson, God exists to serve the human good. Gustafson's theocentrism of course vigorously rejects this. In sum, Gustafson's theocentrism challenges the instrumentalization of God in Kant's ethics, the strong bifurcation of nature and freedom, and its anthropocentric focus.

Though Gustafson ranges through the broad history of Christian theological ethics, he selects Thomas Aquinas and Karl Barth as his primary interlocutors. The reason for these selections is that Aquinas and Barth represent for Gustafson two of the most thorough theological ethical visions within the Christian tradition. Of Aquinas, Gustafson writes that he "has provided the most comprehensive account of theological ethics in the Roman Catholic tradition, and the most systematically developed account in the history of Christian thought."[19] He holds

---

18. Ibid., 136.
19. Ibid., 26.

Barth in a similarly vaulted position with respect to the Protestant theological tradition. Given the prominent place of these great thinkers' systems within the history of Christianity, they are the obvious candidates in relation to which Gustafson needs to delineate his own vision. In what follows, I will outline Gustafson's account of the affinities and differences between his own ethics and those of Barth and Aquinas.

Throughout Gustafson's interpretation of Barth, he stresses the way in which theological judgments condition both methodological choices and content in ethics. Gustafson summarizes these judgments of Barth as including an understanding of God ultimately *for* human beings, the fallibility of human nature, a dialogical construal of the divine-human interaction, an optimism regarding human salvation, and a Christocentric biblical hermeneutic emphasizing divine grace. These judgments, as Gustafson interprets them, serve as the fulcrum of Barth's theological ethics. Along with Barth Gustafson affirms the formal principal that the priority in theological ethics is to articulate a coherent doctrine of God. This seemingly obvious point is not always taken as the point of departure for theological ethicists. Gustafson agrees with Barth that *theological* ethics properly begins with a doctrine of God. But Gustafson's account of God, interpreted in chapter four, indicates theological judgments different from Barth's that reflect different methodological moves and ethical content. And yet, "How God is construed," Gustafson writes, and how the world is construed in the light of one's convictions about the ultimate power and the powers that bring life into being, sustain it, and bear down upon it is the "most critical choice made in theological ethics."[20]

While Gustafson affirms the significance of a doctrine of God for theological ethics, he and Barth part ways on a number of crucial points. In distinct contrast to Gustafson, Barth construes the nature of God in agential terms; and his account of God's relations to the world is therefore also profoundly interpersonal. On Gustafson's interpretation, these positions of Barth's delegitimate "appealing to any ordering patterns of life in the world as a basis on which to make moral choices."[21] This is a crucial difference between Barth and Gustafson influenced by and influencing the differences in their theological methodologies.

20. Ibid., 27.
21. Ibid., 32.

Whereas Gustafson's revised Reformed doctrine of God is shaped by and shapes the methodological priority he grants to experience, Barth is well known precisely for rejecting the theological significance of human experiential categories. According to Barth, movement from the fallibility of human experience to an account of God inevitably presents a distorted image of God. For Barth, in place of human experience or the ordering of the natural world, knowledge of God and thus doctrine of God issue from of a Christocentric reading of the Bible that reveals a personal God who interacts dialogically with humans and assures human salvation.

Biblical revelation, for Barth, is the principal authoritative and only reliable source for a doctrine of God. In contrast, the Bible for Gustafson provides an important record of what others have discerned of the divine will, though it is not the sole reliable source for such knowledge. Nature, history, experience, and scientific inquiry of course are other sources. On Gustafson's reading of the biblical material, a more ambiguous understanding of God is present and Barth's "singular Christological interpretation" is "too simple and neat."[22] On biblical grounds Gustafson challenges Barth's Christocentric assurance that God is always for humans. God is presented in many different ways in the Bible on Gustafson's interpretation—as transcendent and immanent within history and nature, at times concerned with humans and yet at times indifferent. The God of the Bible, writes Gustafson, is "a God who is the ground of security and a God who undercuts most of the grounds of security on which we depend."[23]

Barth's rejection of the methodological priority of experience, his strong dependence on what he deems the biblical understanding of God, and his view of the personal quality of God's involvement with the world underlie his provocative claim that "ethics is sin." In saying this, Barth means that ethics generally grants a too prominent place to practical reasoning, to the human capacities to make correct moral judgments, and thus assumes a too exalted account of human nature. For Barth, right and good are determined by God and God alone, and insofar as ethics presumes the human capacity for such determinations, it sinfully substitutes the moral agent for God. While Gustafson's an-

---

22. Ibid., 35.
23. Ibid.

thropology clearly supports Barth's appraisal of human fallibility, his theological ethical vision is not one of a dialogical call and response between God and humanity. It is instead the human task of discerning within the condition of finitude the will of God through the patterns and processes of the world, of perceiving and making evaluative-descriptive judgments about the norms indicated by God's activity in the world for how humans should relate to all things and others. Rather than understanding the life humans ought to live as the life humans are commanded to obey by Barth's gracious biblical God, the moral and pious life for Gustafson is primarily one of discerning through experience an understanding of the world's ultimately divine ordering. In sum, Gustafson writes that while on one hand "Barth's theology contains a theocentric ethics (it is the ruling and commanding sovereign grace of God that orders life), from another perspective his theology sustains [an excessively anthropocentric] focus on the salvation of persons . . . ."[24] Gustafson affirms the first of these convictions of Barth's, that "only God is God," but rejects the second anthropocentric soteriological focus.[25]

As with Barth's theological ethics, Gustafson affirms and critiques aspects of Thomas Aquinas' moral theology. In his effort to delineate the character of his own theocentric ethics, he continues with Thomas to show the way in which theological judgments influence methodological and content choices in ethics. The organizing pattern of Thomas's system, Gustafson reminds, is the pattern of *exitus et reditus*, the view that all proceeds from and returns to God. Along with Barth and Gustafson, Thomas's ethics views norms as objective to moral agents. But in contrast to Barth's divine command ethics, and more in keeping with Gustafson's own position, the theological pattern of *exitus et reditus* allows that the natural world can serve as a significant source in theological ethics. Gustafson acknowledges that this is one of the principal similarities between his own work and Thomas's and a principal difference from Barth's. He writes that while for Barth, "God is related to nature primarily through man," for Thomas "man is, in very important ways, related to God through the ordering of nature."[26] In contrast to Barth's prioritization of biblical revelation, the natural or-

---

24. Ibid., 28.
25. Ibid., 40.
26. Ibid., 44.

der for Thomas is understood as a more reliable source of moral and theological knowledge. Thus Thomas's theological moral epistemology, along with Gustafson's, includes an empirical bent and his accounts of the natural capacities for moral reasoning and judgment have, in Gustafson's judgment, "a theological dignity that [they] cannot have" for Barth.[27]

But the natural order for Thomas is not of course the only medium of God's relation to humans, and while naturally given human capacities can discern the natural moral law, the intervention of divine grace perfects the moral virtues grounded in these capacities and yields the theological virtues of faith, hope, and love. These two points reflect for Gustafson the two-layered character of Thomas's ethical system: "one is the understanding of nature which comes from observation, reflection, and abstraction not only about human activity but also about the relations of human activity to the wider order of nature; the second is the revealed knowledge of God pertaining to the final ends of nature and particularly of proper human activity."[28] While for Barth the ground of ethics is rooted dialogically in God's personal command, for Aquinas it is ontologically rooted in the order of being, illuminated and brought to further specification through biblical revelation. In contrast to both Barth and Aquinas, Gustafson "deliberately [refrains] from using the term 'revelation' to characterize how humans become informed about God on the grounds that what is called revelation is reflection on human experiences in the face of the ultimate power and powers."[29]

While Gustafson's views resonate generally with Thomas's on the matter of the relation between the natural world and moral knowledge, his theological judgments, shaped as they are by contemporary scientific knowledge, lead him to reject several specific features of Thomas's system. First, though situating the human within the context of the natural world, as a part within the whole of nature, for Thomas "in the end it is the salvation of [human] individuals that is the chief purpose of God."[30] Gustafson of course rejects this anthropocentrism. Second, and related to this first critique, Gustafson challenges Thomas's confidence in the

27. Ibid., 49.
28. Ibid., 46.
29. Ibid., 28.
30. Ibid., 54.

teleological perfection of nature. He writes, "I need only note here that in the light of contemporary interpretations of the future of life in this world and of the universe we know, the fulfillment that Thomas affirms cannot be sustained."[31] For Gustafson the dependence of the world and human life on God does not mean that there is a single harmonious end or telos for all things. Thus Gustafson affirms the partial naturalistic grounding of Thomas's ethics but rejects its anthropocentric and teleological character. A third significant point of contrast between Gustafson and Thomas concerns their anthropologies. Thomas holds, according to Gustafson, to a fundamentally unchanging account of the essence of human nature, qualitatively distinct from the nature of other animals. With insights from the natural sciences, especially evolutionary biology, Gustafson challenges the non-developmental character of this anthropology, the distance it assumes between human and other life forms, and the ethical conclusion Thomas draws that lower forms of life exist to serve human purposes. He writes, "the recognition that there is no timeless essence of man," and "of our biological kinship with other animals qualifies . . . both the significance of the distance between us and the idea that because of this distance the more imperfect are created to be in the service of us who are more perfect."[32] And finally, Gustafson critiques the moral hierarchy or "chain of being" structuring Thomas's system. In place of "hierarchy," Gustafson emphasizes the importance of "interdependence" and his correlative participatory anthropology. A long passage illustrates the importance of this to Gustafson:

> While Thomas surely affirms human freedom . . . the capacities of modern technology have enlarged the range of consequences of human choices. It is for this reason that the idea of man as participant becomes important. We participate in patterns of interdependence, and can alter the conditions on which we depend . . . . We have not dominated nature, even though we have controlled more aspects of it than was possible in previous times. But our capacities to intervene into it mean that 'nature' does not provide the kind of blueprint for proper human conduct that is more strongly the case in the ethics of Thomas.[33]

---

31. Ibid., 55.
32. Ibid., 56.
33. Ibid., 58.

The concept of interdependence in contrast to hierarchy serves for Gustafson as a way to draw out the ethical significance of human participation in the world. While dependent, humans impact the natural world in ways unlike any other animal. In the contemporary world, this is only pronounced. Hierarchy presumes a chain of being from lower to higher, with human beings situated between "beast and angel." The axiological character of this hierarchy of Thomas's, in Gustafson's judgment, is too narrowly circumscribed, much too anthropocentric. Interdependence, in contrast, requires a radical enlargement of the whole, totality, or collective good that is the object of moral concern.

The preceding interpretation of Gustafson's basic view of the nature of theological ethics in relation to other major traditions and figures in philosophical and theological ethics has attempted to isolate what he takes to be the central affinities and differences between his own vision and others. Emerging through this interpretation, and reinforcing aspects of my analyses in previous chapters, are Gustafson's emphases on both character and action, the empirical and naturalistic orientation of his methodology, his critique of agential characterizations of God, his critical qualification of the Bible as a source in theological ethics, and his appreciation for nature's patterns and processes as a source of moral knowledge, a field of religious experience, and a crucial element of his moral anthropology. With these insights in hand, it is now possible to move with greater facility into a focused treatment of Gustafson's construal of the theocentric moral life.

Gustafson's Theocentric Ethics

One helpful way to illuminate understanding of any ethical thinker is to ask what serves as his or her basic moral question. For Jonas, this question might be construed as follows: "What does responsible freedom entail in a time of nature's endangerment by the radical expansion of human power?" Gustafson's ethics hinges on a very distinct basic moral question: "What is God enabling and requiring us to be and to do?" These very different questions indicate what Jonas and Gustafson take to be the normative tasks and aims of moral existence, their centers of moral concern, and their distinctive moral anthropologies. For Jonas, the normative aim is to embody freedom responsibly and to ensure its ongoing possibility; and for Gustafson it is to serve or glorify God.

The center of moral concern for Jonas is nature's endangerment by the contemporary moral ambiguity and irresponsibility of human power. For Gustafson, it is the life of piety before God. And while Jonas's question reflects the significance of freedom and purposiveness within his anthropology, Gustafson's highlights finitude and the human relation to or participation with the divine.

In order to appreciate the force of Gustafson's question, some brief interpretation of his terms is required.[34] "God," for Gustafson, as earlier described, refers to the ultimate power ordering the universe. The universe is divinely governed. God is the source and sustenance of all that is, the power bearing down upon and sustaining all. Gustafson's understanding of God, as earlier detailed, rejects anthropomorphic images. The concept "enabling" refers to the idea that the capacities, the possibilities and limits, of human moral agency in some measure precede what agents actually do. These underlying capacities of agency, prior to acts themselves, are given to humans through dynamic natural and cultural processes ultimately ordered by God. Thus the agent is always tethered to natural and cultural conditions and Gustafson's view of the moral agent needs to be understood, along with Jonas's, as a challenge to more exalted or unencumbered accounts of human freedom. Together, the concepts "enabling and requiring" suggest an ethics comprising both duty and what Gustafson refers to as "aspiration," in short, a blending of deontology and consequentialism. Certain ways of being and doing are required of us, and some are to be aimed at. "Us" for Gustafson refers to individuals, communities, and the whole of the human species. The individual is never an isolated actor, but is always interactively involved in the complex patterns and processes of nature and history, responsive to as well as exerting an influence upon them. God enables and requires communities of agents as well as particular actors to be in particular ways and to do particular things. "To be" designates the character, the qualitative nature of individual and corporate actors that God enables and requires. An agent's activity, "doing," proceeds from the character of the agent's being, even while that activity impacts the agent's character. This reflects Gustafson's emphasis on the necessary interrelation of character and action in ethics and the demand to hold both of these elements together in a comprehensive ethics: "Morality pertains to ac-

34. Ibid., 1–2.

tions, individual and collective, as well as to qualities of individuals and states of affairs."[35] In sum, then, Gustafson's basic moral question asks what God, the ultimate ordering power of the universe, makes possible and demands morally of the activity and character of human individuals and communities.

Gustafson's construal of the norm that flows from this basic question states that, "We are to relate ourselves and all things in a manner appropriate to their relations to God."[36] God enables and requires agents to relate to themselves and to other things, to objects of moral consideration, in accordance with their relations to God, the ultimate ordering power of the universe. This imperative raises to priority a distinctive view of moral reasoning as a process of discernment, which I will briefly touch upon here but more fully explore in my coming interpretation of the subjective aspects of Gustafson's ethics. Moral reasoning as discernment is neither a disembodied, abstract calculus of reasons and values, nor an entirely intuitive enterprise. Discernment as Gustafson intends it combines descriptive and evaluative dimensions. To discern is not exhaustively to describe an object, thing, or event, but to discriminate the morally significant features of an object, thing or event in light of its relation to the divine, and what this relation suggests for moral character and action. The discrimination of significance turns on interpretive decisions regarding descriptive and evaluative criteria and judgments about what is salient with respect to the aim of the interpretation.

The activity of discernment only occurs within the conditions of human finitude. It is a mode of practical reasoning that Gustafson understands to be appropriate to participatory, interactive agents in a theocentric world and that is intractably tied to all that is flawed in human beings. The central human flaw for Gustafson, recall, is "contraction." The task of discerning how one ought to be and to do in relation to God is influenced, as described in chapter four, by the depth and scope of one's moral vision, affection, and valuation. Proper discernment of what God is enabling and requiring demands a three-fold correction:

> an alteration and enlargement of vision, which is in part a correction of the flaw of our rational activities; an alteration and enlargement of the 'order of the heart,' which is in part a correc-

---

35. Ibid., 2.
36. Gustafson, *Ethics from a Theocentric Perspective*, 1:327.

tion of the flaws of idolatry and of disordered loves and desires; and different standards for determining proper human being and action as a result of the other corrections, which is in part a correction of the flaw of 'disobedience.'[37]

Such alterations and enlargements as Gustafson calls for here turn principally on the movement from self- to other-regard—ultimately from anthropocentrism to theocentrism. In theocentrism, vision, affection, and valuation are to be profoundly enlarged. The result is a resituation of the human within a frame of moral reference much broader than anthropocentrism. The resulting quest for human meaningfulness and moral integrity is forced to expand its horizons. The issue of what Gustafson understands to be the mechanics of this correction, the maintenance and reconstruction of communities of moral discourse and their symbol systems, will be treated in this chapter's conclusion.

In sum, the basic question and imperative of Gustafson's theocentric ethics should be understood as interpretive tasks calling agents to evaluate and describe, to be and to do in ways appropriate to their own and other's relations to God. The several layers of this task always are embedded within the core human conditions of finitude and dependence, meaning that the moral task is not often unambiguous and that the correction of enlargement is crucial. This basic formulation of theocentric ethics includes several interlocking positions, objective and subjective normative claims supported by and reflexively referring back to the hermeneutical and fundamental dimensions of his thought. My intent in the following is to articulate these positions in order to ground the critical work in my closing section.

One of the tasks of any ethical vision that seeks to be comprehensive is to hold together its objective and subjective features, its ground, justification, and source of moral value and obligation—the good and the right—and the attitude, posture, or form of life correlated to this ground. I will look next at the objective features of Gustafson's theocentric ethics, his affirmation of God as the axiological center of the moral world, and his commitment to the objective grounding of ethics in the nature of things. Following this, I will interpret the subjective aspects of his vision, critically examining how he characterizes the appropriate

---

37. Ibid., 308.

form or mode of moral being and doing and the general ethos of moral existence.

The core claim of Gustafson's ethics, of course, is that *God is the center of value*. This distinguishes his ethics significantly from Jonas's, as well as from many other traditions in the history of philosophical ethics, and even from some dominant strains in the history of Christian theological ethics. Neither individual humans nor the idea of humanity, as for Jonas, is normatively central. Gustafson's theocentric ethical vision is, by definition, decidedly anti-anthropocentric. If humanity is not the axiological center of moral concern, then the significant ethical question is not, in Kantian fashion, what the categorical imperative demands of practical reason, or the universalizability of a maxim of action. Nor is it about which act or principle maximizes utility, or which course of action leads to happiness.

This does not mean, however, that Gustafson rejects the centrality of humans in the ethical enterprise. Ethics is a distinctly human phenomenon. For the level of critical self-awareness and the cognitive and affective capacities required of ethical thinking are uniquely human. Recall that Gustafson's anthropology emphasizes that humans are fundamentally valuational beings. Thus he affirms that humans are the *measurers* of value in the world, but rejects the tendency to picture humans as value's *measure*. Another way of putting this is to say that for Gustafson ethics is unavoidably anthropogenic but emphatically not anthropocentric. The moral life emerges as a distinctively human form of reflective being and doing in a moral world that is much more than human.

Gustafson argues further that not only does theocentrism understand ethics as a distinctly human enterprise, but also that it does not automatically subordinate the human or any other individual to the good of a wider "whole." It does, however, significantly relativize human and individual value. It qualifies human value and interests, putting these under the critical light of God as the center of value, and sometimes demands restraint and sacrifice of human goods for broader, holistic concerns.[38] If God is the center of the moral world, then the fundamental moral question asks what the world's patterns and processes indicate about what God is enabling and requiring moral agents to be and do.

---

38. Gustafson, *Ethics from a Theocentric Perspective*, 2:6.

The general imperative that follows from this, again, is that humans are to be and to act in ways appropriate to their own and others' (including nonhumans and larger wholes) relations to God. This question and imperative put into significant suspense the place of humans in the moral world, though they do not, in Gustafson's view, inevitably or always compromise collective or individual human goods.

Gustafson's fundamental ideas about God, the world, and humans necessarily fund the objective features of his ethical vision. Common human experience points ultimately, for Gustafson, to the fact that there are greater-than-human powers in the interdependent world bearing down upon and sustaining human life in its multiple dimensions. Monotheistic religious experience for Gustafson indicates *a* power, *an* Other working within, behind, and through the ordering powers of the universe. Thus while all persons, according to Gustafson, experience themselves in relation to powers beyond their control, the sense of dependence on *One* power operating in and through reality is a particularly theological construal of this basic experience. Thus the question of what God is enabling and requiring is not *the* general moral question, but the most basic formulation of the moral question from a theocentric perspective.

Gustafson's hermeneutical commitment to the primacy of experience and his revised Reformed theological framework also clearly support his view of God as the center of value. According to Gustafson, experience of the grandeur and danger of nature, and articulated second-order construals of that experience modified through the natural scientific chastening of human symbol systems, suggest that nature does not exist for human purposes, that human conceptions of good may not always conform to the ultimate purposes of God's ordering of the universe, and thus that it is problematic to construe the idea of God in anthropomorphic terms. The linkage between these points funds Gustafson's argument against the anthropomorphic character of much theological discourse about God as well as the anthropocentric theological axiology that often attends such discourse.

Also central to the objective aspect of Gustafson's theory, directly related to his claim that God is the center of value, is that ethics is objectively grounded in the way things ultimately are. The ethical enterprise is concerned with a moral sphere that extends much further beyond subjective human need, preferences, desires, or values. It is rooted in the broadly interdependent patterns of the world—natural, cultural, social, and historical—in which the activity and development of human moral agents occurs. The task of discerning what God, the center of value, is enabling and requiring of agents thus entails looking to the patterns and processes of the world's interdependence, which serve as a "basis, ground, or foundation for ethics from a theocentric perspective."[39]

As earlier described, such discernment is an evaluative-descriptive project. The will of God cannot be discerned directly from the way the world is described, but is to be critically evaluated in light of experience and the evidences of the sciences. The ways of the world and its multiple features "indicate" or "suggest" what God is enabling and requiring of us.[40] The descriptive and prescriptive elements of Gustafson's ethical theory, then, are intimately bound together. And yet in using the terms "indication" and "suggestion" Gustafson circumnavigates the naturalistic fallacy. What is to be done and valued and respected by the moral agent is discerned from a reasoned and felt interpretation of the objective ordering of the world, but not strictly determined by this ordering. Description and the specification of what "is" the case fund but do not dictate what a theocentric ethics prescribes.

A third distinctive objective feature of Gustafson's theocentric ethics concerns the way he construes the moral sphere. A brief rehearsal of the basic features of the dominant metaethical options in moral theory, consequentialist and deontological, will aid in isolating the distinctiveness of Gustafson's understanding of the moral sphere. In consequentialism, the moral domain is characterized by discourse about good and evil and ascriptions of value. The moral domain is comprised of those things that possess value in some sense and by the agents who have the capacities to register and ascribe value. Moral judgments are referenced generally to the consequences or effects of actions in relation to the human good or telos, the telos of communities, or the telos of nature or of history. In deontological ethics, moral judgments reference what "ought" or "ought not" to be done. Attention to the ends of acts and the consequences of acts is downplayed. Whereas consequential-

---

39. Ibid., 7.

40. Gustafson writes, "The 'ways' of 'nature' are indicators of the ways of God, and inferences are drawn from them to aid human beings to discern how they and all things are to be related to each other in a manner appropriate to their relations to God" (*Ethics from a Theocentric Perspective*, 2:8).

ism prioritizes questions about value, worth and the good, deontology privileges questions about obligation, principles and the right. These ethical frameworks answer the questions of what is good and right by referencing, generally, either the consequences or aims of action or the obligations and principles by which one acts, whatever their consequences may be.

In contrast to these approaches, Gustafson's theocentric ethics builds upon and radicalizes his teacher H. Richard Niebuhr's relational value theory, which is concerned not so much with what is good or right as with what is "fitting", and according to which the moral agent is viewed as a "responder".[41] In Niebuhr's analysis, consequentialist ethics operates with a symbol of the agent as *homo faber*, as desiring or aiming for some end, and reasoning practically about how to attain it. Deontological ethics operates with a symbol of the agent as *homo politicus*, as one whose moral task is to embody and express conformity with moral rules, duties, or principles. In contrast to these, responsibility ethics pictures the agent under the symbol of *homo dialogicus*. The agent is responsive to the multiple forces of nature and culture through which God is acting upon him and the moral task is to respond to these forces, and ultimately to God, in a fitting way.[42] For Gustafson, what is ultimately "fitting" is what God enables and requires of participatory actors. In a naturalistic extension of Niebuhr, Gustafson appeals, as I have been showing, much more to what nature's patterns and processes indicate for a "fitting response" than to Niebuhr's general recourse to the idea of God acting in history, drawn predominantly from biblical materials.[43]

While consequentialism construes the moral sphere by way of a value discourse applied to consequences and intentions of human ac-

---

41. See Niebuhr, *Responsible Self*.

42. Niebuhr crystallizes the differences between these ethical theories in the following way: "If we use the value terms then the differences among the three approaches may be indicated by the terms, the *good*, the *right*, and the *fitting*; for teleology is concerned always with the highest good to which it subordinates the right; consistent deontology is concerned with the right, no matter what may happen to our goods; but for the ethics of responsibility the *fitting* action, the one that fits into a total interaction as response and as anticipation of further response, is alone conducive to the good and alone is right" (*Responsible Self*, 60–61).

43. This is in keeping with my earlier note regarding Gustafson's critique of Niebuhr's agential idea of God. See note 7 above.

tions, and deontology construes the moral through the discourse of principles and rules, both ethical theories generally assume anthropocentric answers to the questions of "Good for whom?" and "Right for whom?" The moral sphere in both of these frameworks is traditionally bounded by human concerns. Gustafson's fundamental ideas of God as the center of value, of the human as a relational and valuational participant within the complex historical and natural matrices of an interdependent world, support a nonanthropocentric account of the moral sphere. The power and powers beyond human control, divine and natural and historical, provide the conditions of possibility and the limits of the moral enterprise, and indicate the norms of moral character and behavior. Because Gustafson is committed to the moral relevance of describing the human as a whole, as one whose intellectual, biological, and cultural experiences interpenetrate, the moral and the nonmoral domains are not rigidly separated. The capacities for being moral that any conception of the moral domain presupposes, according to Gustafson, arise through natural and biological processes. The moral domain is expanded by the idea that God is the center of value and is deepened by its grounding in the multiple layers of the world's interdependence. Thus theocentrism qualifies many traditions in ethics by being concerned with a moral sphere whose scope extends beyond human individuals, human communities, or even the whole of the human species, enveloping the whole of the world that the powers or the power beyond human control bears down upon, limits, and sustains. This position on the overlapping of the moral and the nonmoral reinforces Gustafson's position regarding the objective grounding of ethics. For Gustafson, ethics is objectively grounded in a reality external to and much broader than human subjectivity, ultimately in the divine ordering of the universe.

Gustafson's hermeneutical and fundamental commitments yield a challenge then to historically anthropocentric circumscription of the moral. His hermeneutical commitments and fundamental ideas both affirm human experience as multidimensional. Human experience is falsified when abstracted into discrete cognitive, emotive, or aesthetic categories. Humans experience the natural and historical world, themselves, and others as biological, intellectual, and social creatures. While there are different basic domains of experience, such as nature, history and culture, these domains interpenetrate and experience in each elicits the six natural senses interpreted earlier. The most basic senses are those of dependence and interdependence, of being in relation to powers or a power beyond human control that bears down upon, sustains, makes possible, and limits human life in its multiple aspects. The other senses, gratitude, obligation, and the others each relate to these primal senses of dependence and interdependence.

So, Gustafson's affirmation of God rather than humans as the center of value radically enlarges the sphere of moral concern and entails substituting the historically dominant moral questions of what is good or right for humans with the question of what character of being and quality of action is fitting in relation to God. Moral capacities are made possible and limited by conditions beyond human control—they are given rather than chosen, though choice of course impacts how they are exercised. These natural conditions, because they ground moral experience and enable the exercise of morality, are not rigidly opposed to the projects of moral existence. "Rather than perceiving the moral primarily to stand over against the natural and the historical, as an alien force," Gustafson writes, "moral thinking gives direction to our natural impulses and desires as individuals within the context of arenas of life in which we act."[44] As described briefly in my interpretation of Gustafson's understanding of discernment, moral reasoning is not determined by its conditions of possibility but is necessarily related to them. Moral reasoning operates along with, ordering and restricting, our natural human inclinations and does not oppose them. In fact, moral reasoning cannot get off the ground without the enabling biological capacities of human nature. At the same time, however, the moral projects these conditions enable can impact those conditions, reshaping them and thus providing new conditions of moral possibility and limitation. Morality is enabled by but also can alter its conditions of possibility. In sum, the ground of Gustafson's theocentric ethics is that the moral life is rooted in "an objective 'reality' of which human life is a part," rather than a dimension of reality that is constituted by or limited to human subjectivity.[45] From this objective dimension of Gustafson's ethics the mode, posture, or form of theocentric moral subjectivity follows.

44. Gustafson, *Ethics from a Theocentric Perspective*, 2:8.
45. Ibid.

The objective features of Gustafason's theocentric ethics bring its subjective aspects into relief. Most essentially, describing God as the center of value and bringing into the process of moral reasoning its biological conditions as well as its cultural and traditioned character, lead Gustafson to emphasize the significance of *piety* in the moral life. A religious construal of the experience of finitude—of existing within a matrix of forces ultimately borne down upon, limited, and sustained by a single power, God— induces piety as the appropriate moral posture within the world. As explained in the earlier chapter on his fundamental ideas, Gustafson chooses the concept of piety against faith in order to resist the traditional opposition between faith and reason. Piety is defined by Gustafson in dispositional terms, as an attitude and posture toward the world, others, and God—one that includes both affective and cognitive aspects.

He argues that in contrast to many understandings of faith, piety does not imply a cognitive leap to God from the world of experience and knowledge. Instead, piety is induced or evoked by experience in and knowledge about the world, and captures what Gustafson takes to be the religious character of moral experience. Such experience is never merely emotive, but is a complex integration of reason and affectivity. In short, Gustafson presents piety as a more inclusive religious concept than faith, insofar as it comprehends reason, experience, and emotion, and as a richer account of moral experience than those provided by strictly rationalistic models in ethics. Further, the inclusive meaning of piety resists the historical tendency in theology to dichotomize reason and experience on the one hand from scripture and tradition on the other. Piety is a posture toward the world and others and God evoked by the experienced interpenetration of the reflexive feedback between experience, reason, and the influences of scripture and tradition. To reiterate, for Gustafson the emphasis on piety in the moral life by no means excludes the operations of reason in moral existence. It is precisely this conclusion that piety as an alternative to faith seeks to overcome. Piety holds together the affective-emotional and cognitive aspects of a robust account of moral disposition. Rational activity and emotions operate together, and this mutuality is part of the cognitive gain that Gustafson argues his account of piety has over faith.

The centrality of piety in the theocentric moral life means that, "Ethics from a theocentric perspective is religious ethics not only in the sense that ethical thought is grounded in beliefs about God and God's relations to the world, but also in the sense that the human community lives in part by those senses of dependence, gratitude, obligation, remorse, repentance, direction, and hope that are fundamental religious affections."[46] Though basically a religious concept, Gustafson's articulation of piety provides the possibility of bridging and assessing religious and nonreligious moral constructions of the world. This is so in that the basic senses that ground Gustafson's religious construal of piety can be generated by experiences of reality that are nonreligious. On his view, there is a natural piety that can, in the depth of the experience that evokes it, be correlated with religious piety. Though a nonreligious person may describe the world and experience in it in nonreligious terms, it will be intelligible to religious persons insofar as the experiential terms of rough equivalence, the universal senses, underlie that description. The same holds in reverse. Thus part of the gain the concept of piety has, for Gustafson, relates to its more comprehensive interpretive potential in contrast to faith. This gain is provided by piety's rich moral anthropological underpinnings. Piety reflects and endorses the view that "morality is a matter of the heart as well as the mind."[47] Piety comprehends heart and mind, and, in doing so, more adequately correlates to the multiple aspects of moral experience than do concepts that too sharply distinguish emotion and reason.

But this interpretive gain of piety is less significant for Gustafson than its implications for moral being and doing. Foremost among these implications are the appropriateness of humility and self-sacrifice. Piety is not separable for Gustafson from morality, and entails that in some and perhaps many cases, service to God—relating to oneself and other things in ways appropriate to their relations to God—means sacrificing oneself for the good of others, even the individual for the good of the whole. While it remains true for Gustafson that self-respect and duties toward oneself are significant necessary conditions for moral action and moral responsibility toward others, they are far from sufficient for the theocentric moral life. As the disposition appropriate to life before God, theocentric piety commends moral actions that in other ethical visions

---

46. Ibid., 10.
47. Ibid.

may be presented as supererogatory.[48] In arguing for this intimate tie between piety and morality, Gustafson writes that,

> it is a perversion when piety becomes separated from morality as if God were concerned primarily for the salvation of the individual and not the ordering of the creation, and when morality gets separated from piety, as if moral activity were independent of the wellsprings of the human spirit in both their commendable and perverse forms.[49]

Piety is not for Gustafson merely a religious aspect of the moral life, concerned principally with the individual's relation to the divine. It is emphatically not "piousness." Nor is it a religious analogue to a more strictly philosophical account of moral subjectivity. It is instead the basic form of moral existence that correlates to the way things really are from a theocentric perspective and expresses the radical demands such an existence makes upon moral agents.

Gustafson's account of piety is tied to his understanding of practical reasoning as discernment, earlier touched upon. Theocentric moral discernment is a bi-leveled project of practical reason, the theoretical significance of which is that both levels are simultaneously descriptive and evaluative.[50] It is neither strictly cognitive nor strictly emotive. Its first evaluative-descriptive task is to seek an understanding of what God is enabling and requiring agents individually and corporately to be and to do. In seeking this understanding, Gustafson rules out, as earlier discussed, simplistic appeals to scripture and to a timeless moral order. In place of these sources, he appeals predominantly to human experience in an interdependent world. He argues that what can be discerned about divine governance "are the necessary conditions for life to be sustained and developed."[51] Access to these conditions is through critical reflection on the basic dimensions of human experience of self, nature, culture, and history in which the basic religious senses are present. Knowledge of these conditions and what they enable and require, however, will change as scientific knowledge increases. Knowledge of

---

48. Gustafson writes of this, "What are from other ethical standpoints acts of supererogation are not only obligations within theocentric ethics but are seen as actions which follow from piety" (*Ethics from a Theocentric Perspective* 2:22).

49. Ibid., 11.

50. Gustafson, *Ethics from a Theocentric Perspective*, 1:333.

51. Ibid., 339.

these conditions also is relative to historical and cultural location. Yet, Gustafson claims, these subjective limitations do not mean that theocentric moral discernment is relativistic—for one of the basic assumptions of the theocentric construal is that there is an objective ordering of life. So experience is at the base of discernment, and yet no moral situation, according to Gustafson, is experienced in an unmediated way. This is why discernment is never purely descriptive.[52] The experience that underlies the task of discernment is influenced, for example, by the agent's sense of self, the complexity of his nature as a biological and cultural being, the tradition or traditions in which he has been formed, and by the multiple intersecting spheres of his life experiences. What is being aimed for at this initial level of discernment is an evaluative description of what experience indicates about what God is enabling and requiring.

The second step entails moving from this initial evaluative description to a normative understanding; in other words, discerning what the situation entails with respect to the imperative to relate to oneself and others—human and nonhuman—in ways that are appropriate to one's own and others' relations to God. This level of discernment, too, is influenced by multiple factors. The way one is to relate to oneself and others turns on how one perceives the divine interaction with the world, and such perception is conditioned by one's fundamental ideas of God, world, and human life. Thus the descriptive evaluation of a situation, the understanding of what God is enabling and requiring, and the prescribed course of action and mode of being that follows such an understanding, are never unmediated. As claimed earlier, ethics from a theocentric perspective is a radically hermeneutical project.

Understood in these ways, Gustafson's account of discernment significantly contrasts with other theories of practical reasoning.[53] It is

---

52. My understanding of the reflexive relation between description and prescription, articulated in this chapter's introduction, has been deeply shaped by Gustafson's own.

53. Robert O. Johann goes as far as arguing that Gustafson's view of discernment is really what is most revolutionary about his theocentric ethics. It presents an understanding of practical reasoning that cuts between the polarized and polarizing options of objectivist and subjectivist moral epistemologies. On the one hand, against the tendency of objectivists, Gustafson's view of discernment views practical reasoning as much more than the cognitive enterprise of discovering principles or standards of conduct rooted in some antecedent moral order. As earlier discussed, this is partly

not at all a casuistic process, for example. It does not begin with a catalog of abstract moral principles and then analyze particular moral cases against these in order to deduce what ought to be done. Discernment is much less systematic or formulaic than this. But neither is it a purely intuitive process. It does not appeal simply to intuitions of the good or the right. Rather, it is a deeply reflexive enterprise that includes rational-cognitive and emotive-affective elements. It includes within itself the demand that critical attention be applied to the multiple factors influencing it. It contrasts, obviously, with accounts of practical reasoning regulated by context-independent rules and principles. Discernment acknowledges the particularity of contexts and the way that evaluative-descriptions of contexts are shaped by multiple variables, for example, the scale of context considered, the agent's life experiences, the prevalent symbolic resources of the agent's tradition, and whether the powers of agency exercised are those of an individual or a community.

Another distinctive subjective feature of Gustafson's theocentric ethics, related to those discussed thus far, concerns his emphasis on the inevitably ambiguous and often tragic nature of the moral life.[54] If God is the center of value rather than humans, if the moral and nonmoral domains interpenetrate, if piety best describes the experience of being a moral agent and blends emphases on heart and mind, and if moral reasoning is construed as discernment rather than a rational formulaic enterprise, then the theocentric moral life cannot be a neat and tidy affair.

Gustafson's ethical vision entails a temporal and spatial expansion of the moral horizon. By inflating the frame of moral reference, the relevant descriptive context of theocentric ethics is radically enlarged and complicated. Gustafson's account of the moral agent as interactive participant is crucial here. The theocentric mode of moral being is not that of being a spectator. Along the lines of H. R. Niebuhr, Gustafson

---

signaled in Gustafson's rejection of "order" in favor of the idea of "ordering," to which he calls attention to the dynamic patterns and processes of the world. On the other hand, against radical subjectivists, Gustafson does not view practical reason as an affair of arbitrary choice. Discernment for Gustafson is not entirely subjective, ruled only by inclination. It is instead, as Johann insightfully interprets it, a cognitive-affective process of seeking to make sense of and to organize the whole of experience. Through this subjective process an objective ordering emerges and indicates what one should be and do. See Johann, "An Ethics of Emergent Order," 95–114.

54. See note 5 above.

affirms that the moral task of discerning participation is thoroughly interpretive. Moral discernment for the participant requires a difficult balance of attention to both wholes and parts. Gustafson writes on this score, "If the divine power and powers are, in some sense, providing the conditions of possibility of human action and the ultimate ordering of the creation, the larger context in which individual acts and events take place must be taken into account in morality."[55] This means that while "Individuals (including social collectives as 'individuals') are the agents, [when viewed] in the context of interactions in larger wholes there is a qualification of their discreet acts."[56] If this larger frame of moral interaction qualifies the significance of individual acts, so also does it qualify the work of discerning what participants are enabled and required to be and do. The moral discernment of interactive participants is always immediately tied to discreet circumstances, but this attention to present and local circumstances is mediated and shaped by the whole context in which it occurs, and it is the task of discernment to interpret what this contextualization entails for moral character and activity.

There are of course various ways of construing the morally significant whole with which moral reasoning and moral behavior should be concerned. For some thinkers, the whole is circumscribed by only human life, for others, by all that is sentient, for others still, by all that is living. And with each of these different ways of construing the moral whole, the questions of whether the whole exists for the part or the part for the whole can be answered in various ways. Gustafson's theocentrism answers the first question about the morally significant circumference of the whole with "divine purposes," comprised of the patterns and processes of nature and history, and the second question with the reply that the whole generally has moral priority over the parts—human and otherwise. He writes, "The theocentric construal of reality at least presses us to expand the scope of the wholes that are taken into account in any normative proposal for human action . . . [and that this enlarged scope of moral consideration] greatly complicates ethics . . . ."[57] The essence of complication here is that the expanded context of morality implies ultimately that greater moral significance should be granted to

---

55. Gustafson, *Ethics from a Theocentric Perspective*, 2:12.
56. Ibid.
57. Ibid., 17.

the whole over its parts: "one cannot argue that [the whole] is *only* or even *ultimately* for the sake of our part."[58] Again, if God is normatively central, then the proper aim of the moral life is not to secure human goods, to seek human happiness, or to maximize human utility, but is instead to serve and glorify God's purposes.

In sum, discerning what God enables and requires entails interpreting the patterns and processes as indicators of divine purposes. There is no algorithm for this. There are no pre-established principles or norms that always apply since the world is governed by a divine ordering that is processive and not always harmonious with human goods. Not only does Gustafson's account of moral reasoning resist simple answers, easy methods, or formulaic principles, it unambiguously advances a view of the moral life as inherently ambiguous. For while the whole does not always exist for the sake of the part, it is also the case that precise knowledge of the good of the whole is not easily discernible. The theocentric task is to glorify God, and what this means in particular circumstances, on Gustafson's account, requires a highly developed posture of piety. At the core of piety is the recognition of dependence and finitude, and thus a stance of significant moral uncertainty. The moral aspirations and obligations of participatory agents, in relation to the center of value—the whole of divine purposes—are generally shrouded in ambiguity. Further, these aspirations and obligations may often seem tragic, at least from human points of view. In Gustafson's vision, the common good or the good of the whole extends beyond human social and political spheres to include the patterns and processes of the whole of the natural historical world. This yields potential conflicts not only between individuals and wholes, but also between humans and nonhuman individuals and systems. Whether referencing strictly human social communities or more broadly the ecosphere, the good of the whole and the good of parts are related. But they are not for Gustafson always harmonious.

So for Gustafson, theocentric moral participation is profoundly conflicted and discernment is the densely layered interpretive project of piety. At the same time that the ultimate task is to glorify God, to serve the whole of divine purposes, the pious moral participant is also called to relate to particular parts or individuals within that whole in

---

58. Ibid. (italics original).

ways that honor those parts' relation to God. This is so insofar as the distinctive capacities of particular others, for example the capacity for self-determination in human others, arise through and are part of the dynamic processes of the whole of nature and culture supported and maintained by God. Thus what it means to glorify God and the good of the whole is to be worked out within the specificities of discrete moments and places and relations. But the agent's moral attention needs to be trained on the vast circumference of the whole of the moral world as well as these specificities.

This is an exacting hermeneutical task. That the ultimate concern is the good of the whole, and that the norm for our intra-whole interactions is the relation of parts to God, means "that theocentric ethics will be weighted more readily, in circumstances of conflict between the claims of parts and the whole, on claims for the common good of the whole."[59] Gustafson is well aware of the challenges his holistic vision presents to many cherished philosophical, political, and religious commitments—individual freedom, human rights, resistance to tyranny, for example. And yet for Gustafson, genuinely *theo*centric piety demands, on the basis of what he determines it is possible to know of God through experience, the sciences, and selected theological commitments, that moral priority in cases of conflict be granted to the whole rather than the individual.

When brought together, then, the various objective and subjective features of Gustafson's theocentric ethics present an enormously demanding portrait of moral and religious life. This is seen most emphatically in the stress he places on self-denial: "A theocentric piety . . . motivates and issues in a readiness to restrain particular interests for the sake of other persons, for communities and the larger world."[60] This is not an entirely novel conception, for all moral systems, and the lived experience of the moral ordering of social existence generally, entail prudential balancing of interests between self and others. Some measure of self-restraint is simply required in the effort to get by in the world. This is requisite not only in order to respect the interests, claims, and values of others, whether individual or communal, but also for the interests, claims, and values of selves to have the base conditions neces-

---

59. Ibid., 19.
60. Ibid., 22.

sary for their existence. If there is no restraint at all, then there is only a Hobbesian war of each against every other, and everyone loses.

What distinguishes Gustafson's vision from others on this score, as earlier noted, is that what might be construed "from other ethical standpoints [as] acts of supererogation are not only obligations within theocentric ethics but are seen as actions which follow from piety."[61] And, further, this pious self-restraint is required, for Gustafson, not only with respect to the human social sphere, but also with respect to the natural world of nonhuman individuals and communities. In this way, Gustafson's account of self-denial and restraint can be understood as a radicalization of the biblical tradition's other-regarding demands. And, as indicated by the life of Jesus, for Gustafson the incarnate model of theocentric piety and fidelity, the demand of self-restraint can ultimately entail self-sacrifice: "The cross, and the way of the cross, are revealing symbols of what is enabled and required of persons who seek to serve and glorify God."[62] For neither our own lives nor the world are our own but are given, sustained, and limited by God, the sovereign ordering power of the patterns and processes in which we find ourselves. To be a pious, discerning participant within the given world implies "a readiness to restrain, deny, and even sacrifice justifiable interests of our own for the sake of others."[63] Though humans are not more than parts of the wholes in which they participate, humans according to Gustafson have the capacities as parts to see and feel themselves as able constructively to participate within the wider ends and purposes of the larger whole—even to the point of ultimate personal cost.

Having now interpreted Gustafson in comparative perspective, exegeted his basic moral question and norm, and articulated the objective and subjective aspects of his theocentric ethics, it is time now to close this chapter in an appreciative, but critical light. The aim of this closing is to set the stage for my concluding chapter's tasks of isolating some suggestive ways in which Jonas's and Gustafson's anthropologies complement one another and to recommending responsible participation as a constructive possibility for further development.

61. Ibid.
62. Ibid.
63. Ibid., 290.

Conclusion

In ch. 4 I brought attention to the idea of "intersection" that Gustafson uses to describe his understanding of interdisciplinary work. All phenomena are like intersections, which can be approached from multiple directions. Looked at from one direction, one sees one thing; from another direction, something a little different. For Gustafson, the various directions of approach are the various disciplines, each revealing a distinctive perspective on the intersection. There is a great deal of traffic in Gustafson's work. As this and the previous chapters have shown, Gustafson draws from various thinkers and traditions in philosophy and theology as well as being significantly influenced by the social and natural sciences. Gustafson's is a capacious, erudite mind, thick with traffic like Chicago's Lake Shore Drive at rush hour. All of this converges for Gustafson on one major intersection, the focus of this chapter—the meaning of theocentrism for ethics. The question that now needs to be raised concerns the clarity of the signs at this intersection. Do Gustafson's anthropology of participation and theocentric norm point in a direction appropriate to the human moral and religious journey in a time of radical human power and environmental crisis? Or, differently, do they provide the kind of compass helpful in navigating the tangled bank of human moral life in relation to the peculiar precariousness of life's future?

As I have indicated previously, while there are many who question Gustafson's theocentric ethics, they fasten on a different approach to this question. Some are concerned with how precisely his theocentric vision differs from an ethical naturalism.[64] In addition, some question

---

64. For Gordon Kaufman, Gustafson's ethics is not as clearly theocentric as he suggests. As I earlier quoted, Kaufman argues that Gustafson's conception of God is insufficiently distinct from a more general account of nature. In light of this assessment, "Gustafson's unwillingness to use a straightforward ecological naturalism as the basic framework for his ethics seems even more difficult to understand." See Kaufman, "How is God to Be Understood in a Theocentric Ethics," 27. And as Lisa Sowle Cahill writes, "even as Gustafson's commitment to Christianity diminishes [through the corpus of his life's work] his commitment to theism remains, and his insistence on 'theocentrism' increases." And yet still, she asks, "What or Who is the 'God' of Gustafson's Theism?" See Cahill, "Consent in a Time of Affliction," 31. As I argued in ch. 4, I think these lines of critique do not fully appreciate the multiple sources from which Gustafson's construal of God draws. Gustafson's construal of God is not controlled by nature's patterns and processes, though it treats natural ordering as a significant source. Gustafson's conception of God also draws from other spheres of human experience (culture, history,

the "Christian" character of Gustafson's project.[65] Though I agree that there is room for critique of Gustafson's work, my concern is less with the theological identity of his project, which I think is quite clearly a christomorphically structured revised Reformed theocentrism, than with the sufficiency of his guiding metaphor of participation in a time of pressing environmental challenges.

This line of questioning is invited by the open, reformulative character of the whole of Gustafson's project, as well as by its ethical concern for the whole context of the world's divine ordering. While his theocentric ethics is not limited to concern for the natural environment, he understands the task of discernment forming the life of piety before God to entail evaluative-description of what the world's patterns and processes indicate in relation to the imperative to relate to oneself and others—human and nonhuman—in ways that are appropriate to one's own and others' relations to God. He takes the good of the whole, ultimately the purposes of divine ordering, to have moral priority over any individual, community, or species goods. The question I am pursuing is whether his anthropology and norm of pious participation are sufficient to the task of "relating appropriately to God" in a time in which an evaluative-description of the world's circumstances indicate that the ongoing possibility of the whole of the biosphere is threatened by the extraordinary efficacy of human power.

---

society) and very profoundly from the resources of the Reformed theological tradition. While Kaufman's and Cahill's questions are certainly important to consider, I hold that they arise at least in part from a misunderstanding of the nature of Gustafson's project to develop an account of the religious and moral life from a *theocentric* perspective, christomorphically structured, informed by a deep appreciation for the dynamism of religious traditions, a profound respect for the Reformed theology that has influenced him, and a rigorous engagement with the natural sciences.

65. For Richard McCormick, the fading of Gustafson's vision into "an impersonal deism" turns principally on his low Christology, one that leaves Gustafson "merely thinking about Christ". See McCormick, "Gustafson's God," 64. According to Jeffrey Stout, "[Gustafson] has . . . moved beyond the point at which reformulation tips the balance from a presumption in favor of the Christian tradition to a presumption against it." See Stout, *Ethics after Babel*, 185. As I indicated in ch. 4, such critiques of Gustafson's theology are based to some degree on the privileging of a christocentric over a christomorphic theological vision. To suggest that Gustafson "merely thinks about Christ," as McCormick does, overlooks the way in which Gustafson understands the meaning of being Christian as the project of "continu[ing] to be empowered, sustained, renewed, informed, and judged by Jesus' incarnation of theocentric piety and fidelity" (Gustafson, *Ethics from a Theocentric Perspective*, 1:277).

As claimed at the beginning of this chapter, and as should be evident throughout the others, Gustafson is a highly conscientious methodologist. He holds that Christian theology has always been a dynamic tradition, in some degree influenced by its social, political, and cultural circumstances and the increase of knowledge in the sciences. Theological traditions are developmental, like other traditions of knowledge.[66] He writes, "That theologians do select, reformulate, recombine, abandon, and innovate in the light of knowledge drawn from non-biblical sources is simply a matter of fact. The criteria for making judgments about what is proper in this process are matters of ongoing debate."[67]

This appreciation for the dynamic, reformulative nature of the history of Christian theology is reinforced by Gustafson's characterization of the theological task as the inherently interpretive, principally practical work of construing the world in light of a construal of God. With respect to theology's interpretive task, he writes: "Theology is an effort to make sense out of a very broad range of human experiences, to find some meaning in them and for them that enables persons to live and to act in coherent ways. It reaches to the limit-questions of not only human experience, but also our knowledge of nature." Stressing its practical character, he writes: "the basically practical character of the enterprise is clear; theology is testable in part by its consequences for those whose lives are informed by it: by its adequacy in the light of a broad range of human experiences, by the kind of direction it gives to human action . . . ."[68] Here Gustafson upholds the practical criteria of the consequences of action and normative "direction" as fundamental to the judgments that need to be made regarding the adequacy of theological construction and reformulation.

As I argued in ch. 2, the basic moral problem motivating Gustafson's work concerns human moral evaluations and theological conceptions of divine power. This informs the distinctiveness of Gustafson's vision, the emphasis he places on a construal of God's purposes in relation to

---

66. Ibid.
67. Ibid.
68. For this and the previous quote, see Gustafson, *Ethics from a Theocentric Perspective* 1:158, 159. As previously noted, (see ch. 2, note 38), Gustafson draws appreciatively from Julian N. Hartt for the idea of theology as construal. See Hartt, "Encounter and Inference in our Awareness of God."

the world and his strong critiques of anthropocentric theological and ethical instrumentalizations of God. His critique of anthropocentrism is theologically and morally crucial. Within the history of Christian theology, God has been and continues to be manipulated for human purposes, and this is inconsistent with the divine sovereignty many of these theological traditions affirm. In addition, theological anthropocentrism can and has been used to justify the despoliation of the natural environment, the exploitation of nonhuman lives, and the relatively unchecked growth of human power and the human drive to master nature.

Gustafson's account of theocentric piety, shaped by his nonanthropocentric construal of God's purposes in relation to a more-than-human world and expressed through his anthropology of participation, constructively aims at counteracting these morally problematic tendencies of anthropocentrism. But this constructive counteractive concern leads to a moral anthropology that without emendation is not fully adequate to the moral leverage demanded of a theocentric ethics in a time of environmental crisis, and thus does not entirely satisfy Gustafson's own criterion of providing moral direction to human action.

This judgment turns on my view that Gustafson's construal of circumstances does not provide a full picture of the *reversal of dependence* that characterizes the contemporary human relation to the natural world. Gustafson's construal of his fundamental ideas about the world, human life, and God emphasizes human dependence. As I interpreted this in chapter four, good reasons support Gustafson's account of human dependence. However, while it is descriptively accurate to advance an understanding of human life as embedded within, dependent upon, and participating in the various spheres of the divine ordering, (i.e., nature, culture, society), a construal of things that does not fully weigh the moral gravity of this ordering's dependence on human power is morally insufficient. The peculiar ethos of the present is such that the natural world and the futures of human and other forms of life, indeed the whole of the biosphere, depend on the exercise of human power. The very possibility of participation depends upon a sustaining context of conditions, but it is this very context of conditions that is threatened today.

It is true that while his anthropology of participation highlights human finitude and limitation, Gustafson also argues that it affirms and provides normative direction to the human capacities to intervene in the world. He acknowledges that the increase of these capacities through scientific and technological innovation "qualitatively alters the seriousness of our participation . . . ."[69] Yet, while speaking here to the human power to intervene in the world, and the corresponding moral obligations of this power, Gustafson's account of participation tilts toward an emphasis on finitude and dependence. Though he presents the senses of dependence, gratitude, obligation, remorse, possibility, and direction as coequal, dependence always leads the order and seems to receive greatest emphasis in his anthropology.[70] To the extent that this is the case, participation downplays the radical nature of contemporary human capacities to alter the world, the dependency of the present and future world on contemporary human power, and the uniquely human obligations these circumstances evoke. In light of the present *reversal of dependency* Gustafson's anthropology does not sufficiently attend to the degree to which human participation in the world requires coming to terms with the radical human responsibility for the future of life.

And yet as a metaphor for human being, participation certainly has radical implications for both human knowing and doing—what can be known about God, the world, and ourselves, and what should be done in relation to God, the world, and ourselves. It provides a rich understanding of human life in a more-than-human, interdependent world, and prescribes a norm of caution that is crucial to embrace in a time of environmental crisis. There is great potential in the metaphor to curb the human proclivity toward grandiosity, which is especially dangerous in a time of rampant technological innovation, increasingly intensive interventions in the world, and environmental endangerment.

---

69. Gustafson, *Ethics from a Theocentric Perspective*, 2:281.

70. It is interesting to think about how Jonas might consider some of these senses, and the order of priority in which he might present them. There are several ways of speculating about this. In terms of the scale of life's complexity, "possibility," I think, would lead the order, referring to life's native grounds of freedom, the horizon of transcendence correlated to the various levels organic metabolic mediation. Then might come "direction," the purposiveness innate to life, the will-to-be against not-being. "Obligation" would follow these as uniquely human, the effect and burden of the heightened magnitude of the levels of possibility and direction in human beings. In terms of a moral prioritization, Jonas's emphasis would I think quite clearly be on "obligation" rather than "dependence." The sense of obligation would be understood as the greatest accomplishment of nature, and the obligation to ensure its possibility and continued directive influence would be privileged.

I take this possibility to be one of the great strengths of Gustafson's anthropology.

It is because humans are participants—historically situated, grounded in the patterns and processes of nature, limited and finite creatures—that so little of God's purposes for humans and for the world can be known. It is because humans are participants that we ought not to presume with too much confidence what God is really like. It is because humans are participants that discernment of the possibilities and obligations of the moral life in relation to God are not often obvious. And yet, while each of these understandings is rooted in a richly rendered account of human life in a more-than-human world, and issues from an admirable humility in the face of human limitation, they can yield a moral agnosticism insufficient to the resolve that needs to be summoned in a time for nature and human life as vulnerable as this present time. The tilt toward dependence within Gustafson's construal of participation can underwrite complacency, perhaps even moral paralysis—dangerous possibilities given the stakes and urgency of the environmental crisis for life's future and for the whole of divine purposes that Gustafson's vision of piety rightfully, persuasively urges us to honor.

The anthropology of participation, in Gustafson's words, "sounds a warning signal" and forces us to "take seriously the condition of human finitude with reference to our knowledge, our capacities to predict the effects of our activity, and our capacities to control the interventions we make into the lives of other persons, into social processes and orderings, into historical events, and into nature itself."[71] In light of the seriousness of present circumstances—the reversal of dependence indicated by the vulnerability of the biosphere to human interventions—the warning signal sounded by the concept of participation beckons as well for supplementation. Gustafson's methodological sensitivity to historical context and his concern that theology proceed from an accurate account of its circumstances, his understanding of theological ethical construction as a task of "construal," and his affirmation of the practical aspects of theological construction suggest the importance of continuing to develop the significance of his ethics from a theocentric perspective by inquiring into the ways in which Jonas's norm and anthropology of responsibility may constructively complement it.

---

71. Ibid., 280.

Conclusion

Toward an Ecotheological Ethics of Responsible Participation

Introduction

AS I OPENED THIS BOOK WITH REFLECTIONS ON THE IMAGE OF CHARLES Darwin's "tangled bank," so I will conclude it. I need to emphasize that part of the value of this image for me is that it lends pictorial valence to my conviction that the moral problems associated with the environmental crisis do not have easy solutions. I agree with Jonas and Gustafson that moral existence in our time is both more complex and more grave than ever before—we possess greater capacities to intervene in the natural and social worlds and these interventions produce new moral situations and circumstances that demand rigorous ongoing moral thinking and discipline. My task in this book has not been to untangle the bank of moral life, but to think carefully along with Jonas and Gustafson about the inexorably complex conditions and burdens of contemporary moral life. I take Jonas and Gustafson to be two exemplary guides, each of them providing deeply creative normative proposals in response to the environmental crisis.

In my opening comments, I used the image of the "tangled bank" to express the relationships between the problems to which I wanted to respond. I discussed it as an illustration of the interwoven character of environmental crises; I recommended it as a way to picture the layered rhetorical structure of this book; I interpreted it as an evocative image of the interdisciplinary character of Jonas's and Gustafson's ethical projects. Having examined these constitutive interdependencies of the tangled bank, it is now important to consolidate and advance what has been learned.

In my introduction I argued that the environmental crisis should be understood as a crisis *of* and *for* human power. I meant by this that environmental degradation is largely the result *of* the intensification of human interventions into nature and that the ecologically perilous results of these interventions compromise the earth as habitat. I also indicated the ways in which human power to alter nature has become, through biotechnologies and genetic sciences, a power to alter the nature of human being. Referring to Jürgen Habermas among others, I noted the way in which human power has led to the *dedifferentiation* of the natural environment and the human condition. I also claimed that the proper response to this crisis *of* power was to articulate a moral framework *for* human power. I argued that the environmental crisis demands a moral anthropology, a description of and prescription for the human relation to the natural world, sufficient both to the scope of contemporary human power and to the depth of the natural world's (and therefore human) endangerment. I suggested that it would be crucial, for reasons of descriptive and moral adequacy, to balance concern for human continuity within natural processes with attention to distinctive human capacities. What is required is a way to hold in view the ineluctably tangled kinship of humans with all other living forms and the unique responsibilities of human participants within the natural world. Within a broader ecotheological ethics, such a moral anthropology centers upon the dialectical character of human moral existence in relation to nature.

In light of these claims, I have attempted a thorough elucidation of Jonas's and Gustafson's projects, structured through critical analyses of the basic dimensions of their ethics, centering on their moral anthropologies, advancing what I take to be their greatest insights and isolating their flaws. I have focused on these thinkers specifically because I take them to instantiate my concern that theorizing moral anthropology is at the center of the task of environmental ethics. My methodological thesis has been that each thinker's moral anthropology provides a helpful lens through which to gain understanding of the other aspects of their work. I do not hold that their moral anthropologies determine everything else in their ethics, but that a great deal of their ethical visions can be brought into view in light of their moral anthropologies. Achieving greater insight into the significance of Jonas and Gustafson for environmental ethics has been one of my aims. Another, more constructive aim has been to suggest ways constructively to build upon their insights for the purpose of developing a basis for further work toward an ecotheological ethics of responsible participation.

An ecotheological ethics of responsible participation obviously draws upon the deepest metaphors of Jonas's and Gustafson's accounts of the human relation to nature, responsibility and participation. As such, it does not take these metaphors to be incompatible, though their merits as evaluative descriptions for each thinker are supported by distinctive methodological approaches and yield different normative visions.

Jonas arrives at his understanding of responsibility through an emphatically non-theological, phenomenological-existential speculation on biological fact. The genesis of his anthropology and its correlate ethics is in an interpretation of what reflection on human experience indicates about the character and objective value of the broader world of natural life. What this ultimately reveals for Jonas is the purposiveness and ontologically grounded goodness of the phenomenon of life and the need for human moral responsibility to be future-oriented. Responsibility arises as the central aspect of human existence for Jonas through the long history of evolutionary processes, and obligates humans in the present to preserve its ongoing possibility. Ensuring the future conditions of human responsibility is imperative for Jonas, not for the sake of future human life alone, but for the future of the whole of the natural world and out of dutiful moral regard for the history of natural processes through which responsibility has emerged. Human existence is justified, in spite of its great potential for harm, as nature's own greatest achievement.

Quite differently, Gustafson's anthropology and norm of participation are developed by way of a reflexive scientific and theocentric correlation with human experience. His norm calls for humans individually and collectively to "relate ourselves and all things in ways appropriate to their and our relation to God." Discerning this entails interpreting what the patterns and processes of the various spheres of experience indicate about what God is enabling and requiring humans to be and to do.

As I have previously suggested, Jonas's moral anthropology subtly leans toward human discontinuity with nature, and Gustafson's toward continuity. My claim, however, is not only that there is compatibility in spite of these differences, but that in light of the gravity of the environmental crisis and the complexity of human moral existence—motivational sources for both Jonas and Gustafson—it is helpful to think through the ways in which the metaphors of responsibility and participation can constructively inform one another. Each thinker's metaphor emphasizes a crucial aspect of the character of human moral existence. Phenomenological reflection on moral experience, the methodological base from which both Jonas and Gustafson work, indicates that humans are fully creatures who participate within nature and who also exist in a uniquely mediated relation to nature.

Joining the metaphors of responsibility and participation aims at this dialectical character of the human relation to nature, advancing the insights of each thinker. It is a constructive step to take in response to both Jonas's ultimate concern that the ambiguity of human power poses a grave threat to the future of human life and Gustafson's ultimate concern that the gravest threat is to the whole of the world's divine ordering. These concerns neither need nor should be opposed. Drawing together the concerns and metaphors of Jonas and Gustafson is a constructive task for an ecotheological ethics because they reflect equally important perceptions of what the threatened natural world morally requires of us, and equally crucial insights into the relation between humans and nature. Humans are simultaneously participants dependent upon the whole of nature and uniquely responsible for much that is now and will continue to be dependent upon the exercise of human power. The joining of responsibility and participation provides a suggestive base of reference for navigating the moral complexity of the environmental crisis, a crisis simultaneously *of* and *for* human power.

In order to further the constructive possibilities of responsible participation, and prior to providing a preliminary, suggestive outline of it, two final steps need to be taken. First, I will attend to one crucial way in which the embodiment and enactment of Jonas's ethics of responsibility can be complemented by elements from Gustafson's work. Specifically, I will critique Jonas's political vision, which I interpret to constitute a problematic internal tension within his project. Second, I will attend to one of the possibilities within Jonas's work, his speculative theological writings on the *imago dei* motif, for developing Gustafson's anthropology in a direction that I take to be significant in a time of nature's dependence on human power. I do not intend to advance Jonas's work on the *imago dei* as the only or even necessarily the best way to complement Gustafson's anthropology. Instead, I view it as one possibility that could seed further scholarship.[1]

Complementing Jonas

Jonas's conception of the problem of power leads him to advance an account of the moral task as the effort to direct power responsibly. Ultimately for Jonas, as my interpretation of his speculative theological writings convey, responsibility is imagined as flowering as gift from God through nature and as the burden of human life toward life's future. To preserve the intrinsic good intuitively evident in nature against the threat of human power in a technological age makes it imperative, for Jonas, that human power be responsibly directed toward preserving the conditions necessary for future responsible power, and this makes knowledge a moral duty. Ambiguous power requires countervailing restraint by enlightened responsible power.

Jonas possesses a sharp, prophetic sense of the perilous status of nature and the future of life. I believe that this sense for the gravity of our environmental situation needs to be affirmed and strongly encouraged. Yet there is a tension within Jonas's proposed response that needs to be drawn out. Paradoxically, Jonas's effort to articulate a philosophical ethics adequate to the radical capacities of contemporary human action, and specifically his political conclusions, can be critiqued for subtly reinforcing the dilemma of power to which he is responding. Recall that in ch. 5 I discussed Jonas's call for a green socialism, noting that he judges the need-economy of socialism a better model for our time than the profit-economy of capitalism. Even more importantly for Jonas, socialism has the advantage over democratic capitalism of possessing the power to impose the unpopular regulations that the imperative of responsibility entails.

It is hard to disagree with Jonas's judgment that the demands of ecological responsibility will be unpopular. Sacrificing perceived immediate interests for the future common good is and always will be

1. It is important to acknowledge here that Jonas's work with the *imago dei* motif is undertaken speculatively. He does not demonstrate the historical conscientiousness characteristic of Gustafson. Nonetheless, it arises within the only published exchange between Jonas and Gustafson, and I choose to consider it in some detail for its suggestive potential as a supplementation to Gustafson's anthropology and norm of participation.

difficult. But I argue that even so, and even if time is short, Jonas's call for the centralization of political power within an enlightened ecological elite stalls the formation of responsibility upon which, on his own terms, the good of the future depends.

As delineated earlier, Jonas's political ethics is shaped most significantly by his emblematic paradigm of responsibility, the nonreciprocal responsibility of parents for children. He does not make the transition from the paradigm of parental to political responsibility uncritically. He notes the differences between these models of responsibility, and yet emphasizes their common embrace of the concepts of totality, continuity, and futurity. Emphasis of these commonalities, however, glides over a crucial difference. The telos of parental responsibility is to nurture the independent responsibility of the child, to educate and care for the dependent child in such a way that leads to the child's relative autonomy and thus the terminus of the parent's responsibility. The aim of parental responsibility is then, in a way, is its own eventual obsolescence. When Jonas moves from the dynamic of the parent's nonreciprocated responsibility for the dependent child to political responsibility, he prescribes a paternalistic conception of political responsibility that can be critiqued as ensuring, not the fostering of responsible power, but instead the ongoing adolescence of power.[2] To prescribe this is to prescribe dependence, not responsibility.[3]

Thus Jonas's recommendation of political centralization compromises the viability of his ethics of responsibility in the very sphere in which it is most crucially needed. The norm of responsible power needs to gain traction among the many individual and communal agents who, for reasons that Jonas himself acknowledges—the cumulative nature of power's incremental extension and the long, still effective history of dualistic conceptions of humanity and the natural world—have become oblivious to the radical effects of their capacities and their complicity in the degradation of the biosphere and the undermining of the condi-

---

2. Richard Bernstein levels a similar critique of Jonas's political vision in his article, "Rethinking Responsibility."

3. In addition to this problem, there is a troubling irony in Jonas's political prescription. Recall that for Jonas knowledge is a correlate of power and responsibility. This argument for the political centralization of power and knowledge within a ruling elite is close in structure, if not in moral intent, to the Nazi fascism that prompted the ethical turn in his work to begin with.

tions for the future of life. While Jonas's emphasis on the significance of nonreciprocal responsibility is intended, importantly, to theorize grounds for duties to the *futures* of humanity and life, this emphasis needs to be complemented by a view of the significance of reciprocal responsibilities if the actual exercise of responsible power is to get off the ground. Gustafson's anthropological metaphor of participation provides a helpful leavening influence here by implying an alternative political vision.

Though Gustafson does not explicitly advance a political vision in the way that Jonas has, one can be discerned within his account of communal moral formation. In his conclusion to *Ethics from a Theocentric Perspective*, he writes that, "From the perspective of the present work the importance of [participation within communities of moral discourse] cannot be overstressed."[4] Integral to the descriptive content of his anthropological metaphor is the irreducibly social-relational, interactive character of agents. This influences his evaluation that, "No single person has the capabilities to be sufficiently informed about factual matters, analyses of processes and patterns of interdependence, and projection of possible consequences of alterative courses of action to be absolutely self-reliant."[5]

But for Gustafson the problem is not merely that no single person possesses sufficient knowledge, it is also, and perhaps most importantly, that participation within community provides a check against the fundamental human flaw of contraction. Given that no individual alone is sufficient to the task of discerning what is entailed of moral participation, and that contemporary circumstances theocentrically construed demand discerning and effective moral participants, Gustafson's vision prescribes the political effectiveness of participation and formation within moral communities. In response to its own demands, moral participation, as I interpret it, enjoins one to be constructively involved in the morally formative reciprocal responsibilities of voluntary communal life.

That this obligation is present in Gustafson's ethics, of course, does not mean that he holds that communal moral inquiry and discernment will yield the right, or from the theocentric perspective, the most pi-

---

4. Gustafson, *Ethics from a Theocentric Perspective*, 2:316.
5. Ibid.

ous moral policy, nor even that it will yield consensus. This does not, however, detract from either the purpose or the effectiveness of communal moral inquiry. The primary purpose of moral communities is not to achieve consensus, but, in Gustafson's words, "to help form the 'consciences' of persons, to educate their rational activity, to enable them to think more clearly and thoroughly about the moral dimensions of aspects of life in the world."[6] In addition to this purpose, the various moral perspectives within and between communities, and the moral arguments among them, enriches public moral consciousness by bringing to view its diversity. Diversity in principle is not a good, but it can provide the occasion for enlarging revision of perspectives, lead to the mutual enrichment of moral thought and practice, and foment the constructive alteration of individual and communal moral policies.

Among the various moral communities—familial, civic, political—Gustafson of course attends especially to religious communities, specifically Christian churches. This would seem a long way from Jonas's political ethics, but Gustafson's reflections here help to draw out what I take to be the latent political wisdom within his project. While there is a temptation among every community to interpret and make judgments in accord with its own moral predispositions, this is a temptation that is especially acute, in Gustafson's judgment, for religious communities. In contrast to this, Jonas appears to be overly optimistic regarding the capacity of an elite to govern responsibly for the good of the common future.[7] Gustafson's concern with this temptation is in keeping with his attention to the flaw of contraction, and specifically with his strong critique of the historic tendencies of religion and theology to manipulate God for human purposes. And yet he acknowledges that religious communities are among the vital moral communities in culture and contribute significantly to the moral formation of their adherents as well as to public moral discourse.

---

6. Ibid., 317.

7. That this is so presents another irony in Jonas's politics, for, as discussed in previous chapters, Jonas is highly critical of utopianism. This indicates that even as he advances a socialist vision, in his view this prescription is one of triage. He does not advance socialism as an absolute prescription, but as a meliorative one. This qualification does not deflect my critique, however, insofar as I hold that even meliorative socialism puts off the fostering of responsible power and its internal demand for knowledge among the broad demographic—that is, all of us—upon which life's future depends.

The interpretations and judgments of religious communities are not always distinctive, but according to Gustafson they are or should be backed by religious and theological reasons. This is precisely where Gustafson leverages his strongest critique of such communities. He writes, "What is reprehensible about a great deal of ecclesiastical moralizing about policy issues is the intellectual and academic flabbiness of most of the 'pronouncements,' whether by church agencies or by individuals."[8] This not only betrays such communities' obligations to rigorous thought, but also undermines the broader constructive moral impact of churches. The public contribution of churches, and more broadly of religious communities, depends on the articulation of theological and religious reasons for distinctive moral perspectives and ways of life and the intellectual and moral disciplines of argumentation and faithful practice supporting those ways of thought and life. This account of the tasks of religious moral communities clearly coheres with Gustafson's metaphor of participation, according to which I have argued that the moral participant is construed as not only in fact interrelated with other communities, but also as obligated to participate morally within public life. The task of theocentric moral participation in the patterns and processes of interdependence entails attending to what the ordering of life in nature and history indicate about the ways of God, and attentive discernment and enactment of what is enabled and required depends on the formative and refining roles of communities of moral discourse.

Thus considered, Gustafson's metaphor of participation provides a crucial complement to Jonas's ethics at what I take to be one of its weakest points. For the effectiveness of Jonas's ethics of responsibility depends not on its rational persuasiveness or a centralized mandate, but on the broad-ranging formation of dispositions, attitudes, and ways of thinking necessary to the enactment of ecologically responsible power. It is unlikely that there will be agents responsible in the way Jonas imagines, either in the present or the future, unless they participate within communities that nurture such responsibility. Responsible power needs to be cultivated, or tutored, through exercised participation within morally formative communities, for such communities can draw together the diversity of knowledge and expertise required to negotiate

---

8. Ibid., 318.

the complexity of our moral life, and can check the raw will-to-power toward which our inwardly curved natures much too often lead.

Complementing Gustafson

In light of the account of the environmental crisis as a crisis *of* and *for* human power, or similarly, as a reversal of dependence in the relation between humans and nature, advancing the significance of Gustafson's theocentric ethics entails complementing his metaphor of participation with Jonas's perspective on responsibility. The significance of this complement reflects an increase of concern for the degree to which present human capacities for altering the world have been profoundly amplified and that, likewise, the human moral burden as participants is also amplified. Humans certainly are participants. But increasing human powers over the world mean that humans are responsible for the patterns and processes that all other present and future forms of life will inhabit. While humans as individuals and as a species *participate within* the present conditions of life, contemporary humanity has become *responsible for* the future conditions of participation.

Gustafson would likely answer that there is already a strong theological ethical account of responsibility within his anthropology, and thus no need for the kind of complement I am suggesting. He would urge that his anthropology and ethics are significantly rooted in the senses of gratitude, obligation, remorse, possibility, and direction, along with dependence. But I have previously made the case that his emphasis seems quite clearly to be on dependence. As a result, his participatory conception of the human accents embeddedness within or continuity with nature's patterns and processes rather than human distinctiveness. I have argued that in doing so it tilts away from the gravity of uniquely human responsibility in a time of nature's vulnerability to human power.

Given the reversal of dependence in the human relation to nature, the continuing significance of Gustafson's theocentric ethics can be aided by supplementing participation with a stronger accent on responsibility. My critical concern is not with the rationale supporting Gustafson's anthropology, but with the way that his insights tend to downplay the degree to which humanity is *the* dominant participant within nature, thus with its sufficiency, as it stands, for guiding further work in contemporary ecotheological ethics.

In some respects, Gustafson anticipates and in fact embraces this concern. Theocentric ethics and the anthropology of participation, he very clearly reinforces, do not admit of either moral simplicity or general rules and principles that apply in all situations. The work of participatory discernment is a deeply interpretive, context sensitive, intellectually and morally demanding enterprise. He argues that though his ethical vision may not entirely satisfy philosophical expectations for a neat theory that serves in all situations, no ethical theory ever meets that expectation anyway.[9] If a theory seems to do so, it has not attended to the density of the moral world.

But to respond to the increasing threat human power poses to the natural environment, Gustafson's participatory anthropology and theocentric ethics can be improved by a supplementary emphasis on the ideas, conveyed so well by Jonas, that one salient aspect of the present world's density is the moral gravity of human power and the urgent need to direct this power responsibly, for the sake of the future of life not to mention Gustafson's own theocentric concern for the whole of the world's divine ordering. This suggestion is not one that entails constructing a tidy ethical system that will resolve all environmental ethical scenarios. To the contrary, it is a suggestion the relevance of which is that it may more accurately reflect the untidiness of the present human relation to the natural world.

The task of drawing out this emphasis on responsibility is invited by Gustafson's openness to revision and his appreciation for the dynamism of moral, religious, and theological traditions. As one possible way to pursue this, I will deal at some length in what follows with the only published exchange between Gustafson and Jonas. In this exchange Jonas indicates how he would construe a theological ethical norm and anthropology of responsibility if he were a theologian, focusing specifically on the *imago dei* tradition. It becomes clear here that Jonas's urgent moral concern for the futures of nature and human life have led him to an appreciation for theology as not only a "luxury of

---

9. Gustafson writes that to embrace his participatory anthropology and theocentric ethics is "to see and feel ourselves in a condition of ambiguity" and that this cannot be resolved "by the construction of an ideal ethical theory that provides almost absolute certainty about ourselves and our actions . . ." (Ibid., 290).

reason" but as a vital aspect of an ethics that aims to be adequate to the moral stakes of the present time. I would like to explore how this may resonate with Gustafson's own concern for the way in which the concept of participation demands revision to the way in which humans can and should think about God and the human good. It is an especially appropriate exchange to consider in light of the fact that it centers on what I take both thinkers to share in common, the contemporary moral significance of descriptions of and prescriptions for the human relation to nature.

With characteristic verve, Jonas, the philosopher, entreats Gustafson, the theologian: "Use your surplus and give us the real article! The skeptical hedgings we will supply, but first we want the full blast of your strongest guns. For we well know that immanent reason . . . has not the last word on ultimate questions of our being and at some point must borrow from sources of light beyond itself, or at least consider their witness."[10]

This is a vigorous call to arms, certainly something of a paradox coming from a thinker who has endeavored the greater part of his career

---

10. "Response to James M. Gustafson," in *The Roots of Ethics: Science, Religion, and Values*, edited by Callahan and Engelhardt, 210. This volume presents selections from four years of research organized by The Hastings Center, with support from the National Endowment for the Humanities, on the relationships of ethics to the sciences. Jonas's response is to an interpretive article contributed by Gustafson entitled, "Theology and Ethics: An Interpretation of the Agenda," in which Gustafson lays out some of the methodological assumptions that will guide his, at that time, pending work in *Ethics from a Theocentric Perspective*. Since his article was intentionally interpretive rather than positional, Gustafson took some issue with Jonas's strident critique. In his "Rejoinder to Jonas," Gustafson writes, in defense of his article, "It is too late in the history of theological thought for an academic theologian not to supply his or her own skeptical hedgings. Like Jonas, I believe religion is *a* source of light; if I did not I would be a sociologist or a philosopher, or be satisfied to do 'religious studies.' Maybe 'the full blast from [one's] strongest guns' comes only when one is preaching and not when one is writing a theological paper for nontheological critics" (*The Roots of Ethics*, 219).

This is a legitimate response, given the occasion of the articles. However, Jonas's charge remains highly perceptive, for it reads from the theological ethical methodology of Gustafson's interpretive article toward some of the concerns that other interpreters of Gustafson raise in relation to his at the time incipient, more positional, normative project. I have elsewhere treated these critiques, (i.e. Kaufman's, Stout's, Cahill's, McCormick's), and argued against them in defense of Gustafson. But I also argue, here and in chapter six, that Gustafson's anthropology requires supplementation, and that Jonas's critique provides a suggestive perspective from which to advance that supplementation.

to present a naturalistically grounded philosophical ethics. But earlier in this same exchange, Jonas indicates the philosophical anxiety from which he makes his entreaty. Speaking of his attempt to provide a non-theological moral justification for human existence, which he holds to be pivotal to an ethics of responsibility to nature, Jonas confesses:

> I have tried hard for myself to establish this simple, elementary, and today so important commitment on pure philosophical grounds (persuasive at least, if not compelling), and am by no means sure that I have succeeded. An axiom perhaps to feeling (who would not shudder at the thought of mankind forfeiting its future?), to reason it is a proposition hard to validate, perhaps altogether beyond its scope. Generally, ethics has something to say only about how men should behave to one another, but not that there ought to be men in the first place. Metaphysics must be called in for that—philosophically precarious at best and always dubious in its purported findings. Religion, to the contrary, leaves no doubt about it. What to instinct and emotion is merely repellent—a vanishing of man by our fault—becomes a sacrilege in the theological view.[11]

Understanding this argument requires, at least, a brief exegesis of what Jonas means by "the theological view." By this he intends the common Jewish and Christian affirmations that God created a good world and that humanity is created in the divine image. Of the first of these, Jonas writes that it "endows [the world] with a claim to affirmation by man . . . . [S]uch as it is, this is neither an absurd nor an indifferent universe, neutral as to value, but harbors value, and man with his value-feeling is not lost in it in cosmic solitude and arbitrary subjectivity."[12] That humanity is not lost in an indifferent, value-neutral cosmos, but stationed within it with a very distinct charge, is brought to force on Jonas's reading of the second affirmation. He writes:

> That [the *imago dei* traditions] signif[y] an awesome charge is evident. In its name, humans could be exhorted 'Be ye holy, *because* I am holy, the Lord, your God.' This is a command, not a description . . . it is a capacity, not a possession, and . . . the capacity sets our duty. While in all other cases, the 'goodness' of creation rests securely in the creature and its acting by nature, in man it rests in the endowment with and for a *potentiality*

11. Jonas, "Response to James M. Gustafson," *The Roots of Ethics*, 209.
12. Ibid., 204.

which has been delivered into his hands. While all other life is innocent, he can be guilty of betraying the image.[13]

The meaning of the *imago dei* tradition, for Jonas, is not primarily descriptive or indicative, but prescriptive and imperative.[14] It prescribes a capacity and potentiality, a duty human life is admonished but not guaranteed to meet. No other creature bears the capacity and duty of this image. But what, for Jonas, is the duty of this capacity? He has said that goodness is secure within the sphere of other-than-human life, and that goodness only becomes precarious through the evolution of human beings, and thus that the magnitude and duty of the *imago dei* arises only with humanity. This parallels his naturalistic philosophical interpretation of life's increasing degrees of mediacy, but is expressed here in theological terms.

I previously interpreted Jonas's philosophical defense of the duty to preserve the *idea* of humanity for the sake of life's future, and traced his expression of this through his speculative theological myth. Here in this article, also speculative, Jonas articulates a duty to preserve not the idea, but the *theological image* of the human. Given the confession earlier quoted regarding his lack of confidence in his philosophical justification for human existence, this turn to theological anthropology is no slight verbal alteration.

Further developing this turn, Jonas carefully exegetes the passages from Genesis 2 in which the dual affirmations of "the theological view" are embedded, drawing attention to the narrative cycle of creation and affirmation. He notes that the existence of goodness among animated, "feeling" kinds is more emphatically endorsed than others. While of plant life, God pronounces that it is good, God blesses animals and empowers them to be generative, to "be fruitful and to multiply." Of human lives, divinely imaged, God grants dominion. Of this dominion, Jonas notes that it qualifies exploitation with the idea of trusteeship: exploitation apart from trusteeship, he interprets, "[is] by the universal law of life's feeding on life, the rule in interspecies relations and involves

---

13. Ibid., 206. emphasis original.
14. For the theological and historical significance of this distinction, see n. [xref] in ch. 5.

individual annihilation habitually, but not, on the whole, wiping out of kinds."[15]

Thus interpreted, the *imago dei*, specified by Jonas with the concept of "trusteeship," addresses the singular danger of irresponsible human dominion in the present—humanity's potential to destroy whole species, their ecosystems, and even the biosphere. Today, "For the first time, one creature has become, through his power and freedom, responsible for all the others."[16] This is precisely the moral problem to which Jonas's philosophical naturalism has attempted to respond. In this article, however, Jonas writes: "It is very difficult, if not impossible, from the premises of a purely secular ethics to derive something like a respect for extrahuman life, or obligations toward nature in general. What it can argue is the *unwisdom* of wanton destruction with its eventual, quite natural retribution on man's own life . . . ."[17]

This negative argument for the unwisdom of unbridled power is what Jonas's philosophical ethics offers in its counseled heuristics of fear, and the imperative *not* to jeopardize the future conditions of responsibility. But here Jonas is suggesting that religion, and specifically a theological image of human trusteeship, can offer a more robust imperative, the positive imperative to respect the whole of nature's created goodness.

This brief excursus into Jonas's foray into theological construction brings to light a creative possibility for bridging Gustafson's insights into participation and Jonas's into responsibility, and advancing those insights for future development. I would like now to suggest how Jonas's use of the image tradition exemplifies a way to complement Gustafson's moral anthropology. Doing this entails considering again Jonas's discussion of "image-making," which he identifies as a basic coordinate within his anthropology.

As I interpreted this in ch. 3, imaging for Jonas is a distinctively human capacity, one of the peculiar marks of human existence, a function of the highly mediated human relation to the natural world. It is a result of the degree to which, through evolutionary processes, human beings have come to stand out from nature in a much more intensified

---

15. Ibid., 205.
16. Ibid.
17. Ibid., emphasis original.

way than any other form of life. All life, for Jonas, exists in a mediated relation to its natural environment. But the degree of mediation increases with the complexity of life, from plant to animal to human. Human self- and world-mediation can be seen to be present incipiently in other animal life, and yet with the human, mediacy reaches a new and more heightened level. This is not the case, for Jonas, because there is something supernatural about human life. Rather, the mediacy that shapes humans as self- and world-mediated, and the human form as responsible, is nature pointing beyond itself.

The images humans make are then, in a significant way, nature striving to transcend itself. In contrast to tool-use, which, because it serves biological needs separates human from other animal life in only a fluid way, Jonas writes that image-making characterizes a creature who "either indulges in the production of useless things, or has purposes beyond the biological, or can pursue the latter in a different way than through the instrumental application of things."[18] In light of this, and by way of the naturally emergent image-making requisites of perceiving and depicting likenesses, of intentionally separating *eidos* or form from material reality, humans have stepped significantly beyond the immediate pressures of biological and cosmic time and into history. Through the long, cumulative historical increase of human powers to intervene in the world, humanity has become responsible for the whole present and future of natural life.

As natural-historical, image-making creatures, humans have "purposes beyond the biological," and among these, I submit, is the purpose of imaginatively pondering, representing, and with Gustafson, construing the world toward the aim of living within it not merely as participants but as responsible participants, as trustees of the divine ordering of things. Images, and specifically theological images, can and should aid humanity in the task of participating responsibly in nature. For nature has become vulnerable to human life, and human history and the resources of our morally formative traditions owe a debt of obligation to their natural conditions of possibility. Species and self-images, the moral anthropologies that fund human being and doing, shape the world. As image-making participants within the world, humans are also morally responsible for the moral images that shape our lives. Image

---

18. Jonas, *Mortality and Morality*, 79.

construction and reconstruction shape and reshape the world. And, of course, this world-shaping image-construction can be undertaken for better or worse.

As I interpret Jonas's exegesis of the *imago dei* motif, specified especially by the moral charge of trusteeship, he provides a suggestive theological image for human being that simultaneously honors and advances his own distinctive insights and Gustafson's. He presents an image of the dialectical character of the human moral relation to nature, comprising emphases on both human responsibility for the natural world and human participation within it.

The accent on responsibility flows from the divine charge entrusting humanity with dominion over the natural world. Humans are accountable for the divine ordering by way of the capacity and duty of the image they bear. At the same time, it articulates the embeddedness of humanity within creation, reinforcing the insight of Gustafson's anthropology of participation. Human responsibility is in part a distinctive duty correlated to a distinctive capacity, but is beholden to ensure the security of what humans share in common with the rest of life and the maintenance of the conditions of life's possibility.

Jonas's speculative construal of the *imago dei* motif, and its specification through the concept of trusteeship, renders in one image the significance of Gustafson's participatory norm to relate to things in a manner appropriate to God, and his own concern for the extraordinary responsibilities of humans within and for the vast panoply of life. But in the end, I only suggest this as one possible way in which Jonas's insights could complement Gustafson's. There is no logically necessary reason to revise Gustafson. Revision is not required for Gustafson's project to be internally coherent. And yet, though not required for the sake of coherence, I hold that it advances the significance of Gustafson's theocentric ethics for an environmentally endangered time. With this in mind, I turn in closing to a preliminary sketch of directions to pursue toward an ecotheological ethics of responsible participation.

Toward Responsible Participation

Work toward an ecotheological ethics of responsible participation is motivated by intense moral concern for the degree to which contemporary human capacities for altering the world have been profoundly amplified and, correlatively, that the moral responsibility of human participation in the world is also amplified. The environmental crisis is an interwoven crisis of and for nature and humanity. It is fundamentally a crisis of the habitability of the earth in which nature has become dependent on humans.

The priority environmental moral problem of the present ethos is the endangerment of life through the destruction of the tangled bank that sustains it. This problem cannot be meliorated by narrowly conceived moral initiatives because it is comprised of a manifold of interlocking issues such as ecological sustainability; environmental, economic, and social justice; the complexity of scientific advances and the increasing capacities of the human species for technological intervention; and the failure of foresight and the moral reluctance of formative social, political, and religious moral communities to recognize and creatively respond to nature's endangerment and thus also human vulnerability. What is required in the face of this vast complex of moral issues is a comprehensive vision adequate to it. At the center of this vision should be a conception of the human that does justice to the dialectical character of human moral experience as natural historical beings, simultaneously responsible for and participants within nature.

Because humans are multidimensional creatures, purposive and relational, cognitive, valuational, and affective beings in whom, through divine creativity, nature has become conscious of and beholden to its value and source, the task of an ecotheological ethics needs to be interdisciplinary. The task is to hold together sensitivity to both the demands of the immediate moral circumstances and the historic and future trajectories that inevitably mediate them. The work of practical reasoning appropriate to moral creatures living within an endangered natural historical world, tangled as it is, will necessarily be interpretive.

While no single source or community is sufficient to developing an ethics of responsible participation, emphasis is placed on theological tradition and religious community. For apart from the historical wisdom of our moral traditions, and in this case theological traditions specifically, neither the sciences nor human experience can fully prescribe what the natural historical circumstances demand of us. Responsible participation is theologically rooted in beliefs in and about the goodness of the whole and each part of the tangled bank of creation, and the dominion of humanity accountable to God's ongoing creative purposes.

Responsible participation indicates the discontinuous continuity of human existence within the natural historical world, and builds upon the gift and obligation of human moral and imaginative capacities to respond to the environmental crisis. Responsible participation presents a normative image of humanity as an image-bearing, image-making project of nature's divinely affirmed processive goodness. Humans certainly are participants *within* the world, the world that is created through the myriad of natural processes of divine ordering, and the world that is made and continually remade through human interventions. But the increasing powers of humanity *over* the world mean that humans are responsible for the patterns and processes in which all other future forms of life participate both now and in the future. While we always swim within the streams of natural and cultural history, responsible participation underscores downstream moral obligations. It is because of the massive threat human power poses to nature's future that, now, in this historical moment, humanity must embrace its enormous responsibility for the ongoing possibility of life's participation in nature.

Responsible participation holds that the image-making capacities of humanity can and often are distorted. The image of humanity is fractured and human image-construction can be fracturing. Humanity is gifted and burdened like no other form of life with powers and capacities to shape the world, and though there is creative potential within these, they are all too often directed in destructive, myopic, irresponsible ways. And given the scale of humanity's present powers, even efforts to aim interventions creatively need to be cautiously undertaken. That this is so, that the image of humanity is fractured, that human capacities outstrip knowledge of consequences, indicates that the obligation to bear the image of responsible participation can never be finalized but is always also an ongoing, rigorously demanding aspiration. The image of responsible participation can only be borne imperfectly within the tangled bank of natural history. Out of concern for nature's vulnerable goodness, and the ambiguous potential of fractured humanity, responsible participation strives to hold together in dialectical embrace the ideas that human life *participates within* and is *responsive to* nature's present conditions and that human life is *responsible for* the future conditions of nature and history within which all others will participate and to which they will respond.

The moral burden of the responsible participant, from a theological ethical perspective, is not only toward the present and future of the natural environment, but also to the creative purposes of God within natural and historical life. Responsible participation affirms that human life participates within and is responsive to the divine ordering of creation, and also that in finitude and dependence human life is beholden to honor the divine gift of responsible participation by ensuring its future role in God's ongoing creative causes for the world. Responsible participation aims to respond to some of the crucial concerns of the contemporary ethos: to the vulnerability and finitude of nature and thus of human life, to the extension of human power through technological innovation, to the natural historical character of human existence, and to the theological depth of experience at which these concerns coalesce. Responsible participation indicates the dialectic of human moral experience by providing an image that reflects the limitations, possibilities, and obligations of human life in a more-than-human world.

The moral paradox of the environmental crisis is that while it is a crisis *of* and *for* human power, largely caused by human interventions into the natural world, and largely demanding normative constraint of these interventions, human moral life is deeply tangled within an interdependent bank of natural and historical patterns and processes, and ultimately the world's divine ordering. The environmental crisis is generated by unbridled human power, and demands a normative framework in which power can be responsibly guided, reformed, and corrected; and at the same time, the task of developing such a framework demands an appreciation for the participatory character of the human moral relation to nature and the contexts of interdependence that simultaneously limit and make responsibility possible.

Responsible participation embraces the normativity of moral responsibility along with the normativity of discerning participation as a constructive tension, holding that the moral force of each metaphor arises most fully in relation to the other. Participation and responsibility mutually inform one another and thrive in a tangled relation that should not be untangled—one cannot fully be a moral participant if one is not responsible, one cannot fully be morally responsible if one is not a participant. Responsibility and participation are two aspects of the dialectic of the human moral relation to nature that should be joined in creative possibility for the sake of the future of life and God's yet unfolding purposes for the world. Further development of an ecotheological ethics of responsible participation can provide a moral context for calibrating the complex discussion of theological ethics in a time in nature and history full of promise and yet weighted with so much peril.

Bibliography

Barbour, Ian. *Religion in an Age of Science*. New York: Harper and Row, 1990.
Beck, Ulrich. *Ecological Enlightenment: Essays on the Politics of the Risk Society*. Translated by Mark A. Ritter. Amherst, NY: Humanity, 1995. Originally published as *Politik in der Risikogesellschaft*. Frankfurt am Main: Sihrkamp Verlag, 1991.
Beckley, Harlan. "A Raft that Floats: Experience, Tradition, and Sciences in Gustafson's Theocentric Ethics." *Zygon: Journal of Religion and Science* 30, no. 1 (1995) 201–9.
Beckley, Harlan, and Charles Swezey, editors. *James M. Gustafson's Theocentric Ethics: Interpretations and Assessments*. Macon, GA: Mercer University Press, 1988.
Bellah, Robert. "Gustafson as Critic of Culture," in *James M. Gustafson's Theocentric Ethics: Interpretations and Associations*, edited by Harlan Beckley and Charles Swezey, 143. Macon, GA: Mercer University Press, 1988.
Bernstein, Richard. "Rethinking Responsibility." *Hastings Center Report* 25, no. 7 (1995) 13–20.
Botkin, Daniel. *Discordant Harmonies: A New Ecology for the 21st Century*. New York: Oxford University Press, 1990.
Cahill, Lisa Sowle. "Consent in a Time of Affliction: The Ethics of a Circumspect Theist." *Journal of Religious Ethics* 13, no. 1 (1985) 22–36.
Callahan, David, and H. Tristram Engelhardt Jr., editors. *The Roots of Ethics: Science, Religion, and Values*. New York: Plenum, 1976.
Darwin, Charles. *The Origin of Species*. New York: Mentor, 1958.
Davaney, Sheila. *Pragmatic Historicism: A Theology for the Twenty-First Century*. Albany: State University of New York Press, 2000.
Donnelly, Strachan. "Bioethical Troubles: Animal Individuals and Human Organisms." *Hastings Center Report* 25, no. 7 (1995 Supplement) 21–30.
———. "Philosophy, Evolutionary Biology, and Ethics: Ernst Mayr and Hans Jonas." *Graduate Faculty Philosophy Journal* 23 (2001) 147–63.
Frymer-Kensky, Tikva, David Novak, Peter Ochs, David Fox Sandmel, and Michael A. Signer. *Christianity in Jewish Terms*. Boulder, CO: Westview, 2000.
Giddens, Anthony. *Runaway World: How Globalization is Reshaping Our Lives*. New York: Routledge, 2003.
Gustafson, James. *Can Ethics be Christian?* Chicago: University of Chicago Press, 1975.
———. *Ethics from a Theocentric Perspective*. 2 Vols. Chicago: University of Chicago Press, 1981 and 1984.
———. *An Examined Faith: The Grace of Self-Doubt*. Minneapolis: Fortress, 2004.
———. *Protestant and Roman Catholic Ethics*. Chicago: University of Chicago Press, 1978.

———. *A Sense of the Divine: The Natural Environment from a Theocentric Perspective.* Cleveland: Pilgrim, 1994.

———. *Treasure in Earthen Vessels.* Chicago: University of Chicago Press, 1961.

Habermas, Jürgen. *The Future of Human Nature.* Oxford: Polity, 2003.

Hartt, Julian N. "Encounter and Inference in our Awareness of God." In *The God Experience,* edited by Joseph P. Whalen, 51–54. New York: Newman, 1971.

Hauerwas, Stanley. "Time and History in Theological Ethics: The Work of James Gustafson." *Journal of Religious Ethics* 13, no. 1 (Spring 1985) 3–21.

Heidegger, Martin. *Basic Writings.* Edited by David Krell. New York: Harper Collins, 1977.

———. *Being and Time.* Translated by John Macquarie and Edward Robinson. New York: Harper, 1962.

Heschel, Abraham. *Who is Man?* Stanford: Stanford University Press, 1965.

Hinze, Christine Firer. *Comprehending Power in Christian Social Ethics.* Atlanta: Scholars, 1995.

Jaki, Stanley L. *The Road of Science and the Ways to God.* Chicago: University of Chicago Press, 1978.

Johann, Robert O. "An Ethics of Emergent Order." In *James M. Gustafson's Theocentric Ethics: Interpretations and Assessments,* edited by Harlan Beckley and Charles Swezer, 95–114. Macon, GA: Mercer University Press, 1988.

Jonas, Hans. *The Gnostic Religion: The Message of the Alien God & The Beginnings of Christianity.* 3rd ed. Boston: Beacon, 2001. Originally published in two volumes as *Gnosis und spätantiker Geist.* Göttingen, 1934 and 1954.

———. *Imperative of Responsibility: In Search of an Ethics for the Technological Age.* Chicago: University of Chicago Press, 1984. Originally published as *Das Prinzip Verantwortung: Versuch einer Ethik für die technologische Zivilisation.* Frankfurt am Main: Insel, 1979.

———. *Mortality and Morality: A Search for the Good after Auschwitz.* Edited by Lawrence Vogel. Evanston: Northwestern University Press, 1996.

———. *The Phenomenon of Life: Toward a Philosophical Biology.* Evanston: Northwestern University Press, 2001. Originally published by Harper and Row, 1966.

———. *Philosophical Essays: From Ancient Creed to Technological Man.* Chicago: University of Chicago Press, 1974.

———. "Toward a Philosophy of Technology." *Hastings Center Report* 9, no. 1 (1979) 34–43.

Kass, Leon. "Appreciating *The Phenomenon of Life.*" *Hastings Center Report* 25 (1995) 3–12.

———. *The Hungry Soul: Eating and the Perfecting of Our Nature.* Chicago: University of Chicago Press, 1994.

Kaufman, Gordon. "How is God to Be Understood in a Theocentric Ethics?" In *James M. Gustafson's Theocentric Ethics,* edited by Harlan Beckley and Charles Swezey, 13–38. Macon, GA: Mercer University Press, 1988

Kranzberg, Melvin, editor. *Ethics in an Age of Pervasive Technology.* Boulder, CO: Westview, 1980.

McCormick, Richard A. "Gustafson's God: Who? What? Where? (ETC.)." *The Journal of Religious Ethics* 13, no. 1 (1985) 53–70.

Melle, Ullriche. "Responsibility and the Crisis of Technological Civilization: A Husserlian Mediation on Hans Jonas." *Human Studies* 21 (1998) 329–46.
Midgley, Mary. *Beast and Man: The Roots of Human Nature.* New York: Routledge, 1995.
———. *The Ethical Primate: Humans, Freedom, and Morality.* New York: Routledge, 1994.
Niebuhr, H. Richard. "The Center of Value." In *Moral Principles of Action*, edited by Ruth Nanda Ashen, 162–75. New York: Harper, 1952.
———. *The Responsible Self.* New York: Harper and Row, 1963.
Niebuhr, Richard R. *Schleiermacher on Christ and Religion: A New Introduction.* New York: Scribner, 1964.
Norton, Bryan G. *Toward Unity Among Environmentalists.* New York: Oxford University Press, 1991.
Peterson, Anna L. *Being Human: Ethics, Environment, and Our Place in the World.* Berkley: University of California Press, 2001.
Reeder, John P. Jr. "The Dependence of Ethics." In *James M. Gustafson's Theocentric Ethics: Interpretations and Assessments*, edited by Harlan Beckley and Charles Swezey, 119–37. Macon, GA: Mercer University Press, 1988.
Schweiker, William. *Power, Value, and Conviction: Theological Ethics in the Postmodern Age.* Cleveland: Pilgrim, 1998.
———. *Responsibility and Christian Ethics.* Cambridge: Cambridge University Press, 1995.
Scott, Peter. *A Political Theology of Nature.* Cambridge: Cambridge University Press, 2003.
Stout, Jeffrey. *Ethics After Babel: The Languages of Morals and their Discontents.* Boston: Beacon, 1988.
Taylor, Charles. *Sources of the Self: The Making of Modern Identity.* Cambridge, MA: Harvard University Press, 1989.
Troeltsch, Ernst. *Writings on Theology and Religion.* Edited and translated by Robert Morgan and Michael Pye. Atlanta: John Knox, 1977.
Toulmin, Stephen. "Nature and Nature's God." *The Journal of Religious Ethics* 13 (1985) 37–52.
Vogel, Lawrence. "Does Environmental Ethics Need a Metaphysical Grounding?" *Hastings Center Report* 25, no. 7 (1995) 30–39.
———. "Hans Jonas's Diagnosis of Nihilism: The Case of Heidegger." *International Journal of Philosophical Studies* 3, no. 1 (1995) 55–72.
———. "Jewish Philosophies After Heidegger: Imagining a Dialogue Between Jonas and Levinas." *Graduate Faculty Philosophy Journal* 23, no. 1 (2001) 119–46.
Whalen, Joseph P., editor. *The God Experience.* New York: Newman, 1971.
Wilson, E. O. *On Human Nature.* Cambridge, MA: Harvard University Press, 1978.
Winner, Langdon. *The Whale and the Reactor: A Search for Limits in an Age of High Technology.* Chicago: University of Chicago Press, 1986.
Wolters, Gereon. "Hans Jonas' Philosophical Biology." *Graduate Faculty Philosophy Journal* 23, no. 1 (2001) 85–98.

www.ingramcontent.com/pod-product-compliance
Lightning Source LLC
Chambersburg PA
CBHW071244230426
43668CB00011B/1586